Lecture Notes in Mechanical Engineering

Lecture Notes in Mechanical Engineering (LNME) publishes the latest developments in Mechanical Engineering—quickly, informally and with high quality. Original research reported in proceedings and post-proceedings represents the core of LNME. Volumes published in LNME embrace all aspects, subfields and new challenges of mechanical engineering. Topics in the series include:

- Engineering Design
- Machinery and Machine Elements
- Mechanical Structures and Stress Analysis
- Automotive Engineering
- Engine Technology
- Aerospace Technology and Astronautics
- Nanotechnology and Microengineering
- Control, Robotics, Mechatronics
- MEMS
- Theoretical and Applied Mechanics
- Dynamical Systems, Control
- Fluid Mechanics
- Engineering Thermodynamics, Heat and Mass Transfer
- Manufacturing
- Precision Engineering, Instrumentation, Measurement
- Materials Engineering
- Tribology and Surface Technology

To submit a proposal or request further information, please contact the Springer Editor of your location:

China: Ms. Ella Zhang at ella.zhang@springer.com
India: Priya Vyas at priya.vyas@springer.com
Rest of Asia, Australia, New Zealand: Swati Meherishi
at swati.meherishi@springer.com
All other countries: Dr. Leontina Di Cecco at Leontina.dicecco@springer.com

To submit a proposal for a monograph, please check our Springer Tracts in Mechanical Engineering at https://link.springer.com/bookseries/11693 or contact Leontina.dicecco@springer.com

Indexed by SCOPUS. All books published in the series are submitted for consideration in Web of Science.

Dhwanil Shukla
Editor

Lighter Than Air Systems

Proceedings of the International Conference on Design and Engineering of Lighter-Than-Air Systems 2022 (DELTAs-2022)

Springer

Editor
Dhwanil Shukla
Department of Aerospace Engineering
Indian Institute of Technology Bombay
Mumbai, India

ISSN 2195-4356 ISSN 2195-4364 (electronic)
Lecture Notes in Mechanical Engineering
ISBN 978-981-19-6048-2 ISBN 978-981-19-6049-9 (eBook)
https://doi.org/10.1007/978-981-19-6049-9

This Springer imprint is published by the registered company Springer Nature Singapore Pte Ltd.
The registered company address is: 152 Beach Road, #21-01/04 Gateway East, Singapore 189721, Singapore

Conference Committee

General Chair:

Dr. Rajkumar S. Pant

Convener:

Dr. Manish Tripathi

Editorial Committee:

Chair

Dr. Dhwanil Shukla

Technical Committee:

Chair

Dr. Manish Tripathi

Members:

Dr. Mohammad Irfan Alam
Dr. Sohan Suvarna
Mr. Shashwat Trivedi

Publications Committee:

Dr. Manikandan M. (Chair)

Members

Dr. Sharad Prabhu
Mr. K. M. Kiran Babu

Outreach Committee:

Mr. Saurabh Bagare (Chair)

Members

Mr. Apurv Tiwari
Mr. Shantanu Gulawani
Mr. Pranav Gupta

Event Management Committee:

Members

Mr. Nawaz Motiwala
Mr. Nouman Uddin
Maj. Rahul Naudiyal
Sqn. Ldr. Salil Sharma

Model Airship Regatta Committee:

Contest Director

Mr. Sohrab Mistri

Reviewers

Dr. Rajkumar S. Pant, Professor
Dr. Anirban Guha, Professor
Dr. Brandon Buerge, Associate Teaching Professor
Dr. Luiz Carlos Góes, Professor
Dr. Barry Prentice, Professor
Mr. K M Kiran Babu, Project Research Scientist
Dr. Manikandan M, Assistant Professor
Dr. Manish Tripathi, Postdoctoral Fellow
Dr. Sharad Prabhu, Project Research Scientist
Dr. Sohan Suvarna, Research Scientist

Preface

This book contains selected peer-reviewed papers from the International Conference on Design and Engineering of Lighter-Than-Air Systems (DELTAs-2022) which was held in Indian Institute of Technology Bombay (IIT Bombay), Mumbai, between June 22 and 24, 2022. The conference was organized by the Lighter-Than-Air Systems Laboratory (LTA Lab), IIT Bombay, in collaboration with The Airship Association, The Institution of Engineers (India), Aeronautical Society of India (AeSI), International Women Professionals in Aviation and Aerospace (IWPA), and IMIEU U-LTA.

The conference's reach was around the globe, with participants and registrants from over ten countries including UK, Japan, France, Italy, Germany, USA, Canada, and India. The international conference was a first-of-its-kind event based in India themed on the Lighter-Than-Air Systems technology, with the aim of bringing out the current research trends, advancement in the field as well as highlighting the need for research in LTA systems for future of green aviation. The papers presented at the conference were themed around a number of key topics related to LTA systems including LTA systems design, aerodynamics, structures, materials, stability and control, operations and ground handling, multidisciplinary design optimization, and novel applications.

The paper presentations over the two days of the conference were divided into four technical sessions, namely:

- Session 1: Aerodynamics
- Session 2: Design and Optimization
- Session 3: Materials and Structures
- Session 4: Stability and Control

The conference also featured several keynote sessions by eminent persons involved in the field of LTA systems including those from academic as well as industry experts. Apart from these, the event hosted a number of paper as well as sponsored poster presentations and two panel discussions.

In total, the conference received 53 extended abstracts submissions out of which 18 were selected for presentation post-rigorous peer review by experts in the relevant fields. Out of the papers presented at the conference, 17 papers have been selected for publication in the present book.

Contents

About the Editor

Dr. Dhwanil Shukla is currently an Assistant Professor at the Department of Aerospace Engineering, Indian Institute of Technology (IIT) Bombay. He obtained his Bachelor of Technology (mechanical engineering) from IIT Gandhinagar, and Master of Science (aerospace engineering) and Ph.D. (aerospace engineering) from the Georgia Institute of Technology, USA. His major areas of research interests include low-speed aerodynamics, rotorcraft aerodynamics, and flow diagnostic techniques. He has published 30 papers in journals and conferences of national and international repute. He worked on various research projects like slung load divergence speed prediction, aerodynamic loads on bluff bodies, dynamic flow field characteristics on retreating rotor blade, and low Reynolds number multi-rotor aerodynamic interactions. Dr. Shukla is a two-time recipient of the Vertical Flight Foundation (VFF) scholarship. He is also an awardee of the President Gold Medal and Institute Gold Medal at IIT Gandhinagar.

Effect of Geometrical Parameters of a Tethered Aerostat on Longitudinal Stability Boundaries

Rakesh Kumar and A. K. Ghosh

Nomenclature

a	Distance along balloon center line from nose to reference point, m
A	Aspect ratio
B	Buoyancy force, N
$\overline{c}$	Mean aerodynamic chord of tail, m
C_{Dc}	Tether cable drag coefficient
C_D, C_L	Drag and lift and coefficients, respectively
C_{m}	Pitching moment coefficient
d_c	Tether cable diameter, m
$D_{\max}, L$	Maximum diameter and length of the aerostat, m
F_X, F_Z	External forces acting on balloon parallel to *x*- and *z*-axes respectively, N
h_{br} or H_{br}	Component of distance from RP to COB, +ve for COB below RP, m
h_{cg} or H_{cg}	Component of distance from RP to COM of balloon, positive for COM below RP, m
h_{sr} or H_{sr}	Component of distance from RP to COM of balloon structure, +ve for COM below RP, m
I_x, I_y, I_z	Rolling, pitching and yawing moments of inertia, respectively, about balloon COM, kg-m^2
I_{xy}, I_{xz}, I_{yz}	Products of inertia in *XY*-, *XZ*- and *YZ*-plane respectively, kg-m^2
k_{xx}, k_{yy}, ik_{zz}	Tether force per unit displacement in *x*-, *y*- and *z*-axis, respectively, at BCP, N/m

R. Kumar (✉)
Aerospace Engineering Department, PEC, Chandigarh, India
e-mail: rakesh@pec.edu.in

A. K. Ghosh
Department of Aerospace Engineering, IIT, Kanpur, India

D. Shukla (ed.), *Lighter Than Air Systems*, Lecture Notes in Mechanical Engineering,
https://doi.org/10.1007/978-981-19-6049-9_1

k_{xz}, k_{zx}	Tether x-force per unit of z-displacement at BCP and vice versa, N/m
$k_{x\theta}$, $k_{z\theta}$	Tether x- and y-force per unit of pitch displacement respectively N/rad
$k_{y\varphi}$, $k_{y\psi}$	Tether y-force per unit of roll and yaw displacement, respectively, N/rad
$k_{\theta x}$, $k_{\theta z}$	Tether pitching moment per unit of x- and y-displacement, N-m/m
$k_{\theta\theta}$, $k_{\varphi\varphi}$, $k_{\psi\psi}$	Total tether pitch, roll and yaw moment per unit of pitch, roll and yaw displacement, respectively, about COM, N-m/rad
$k_{\theta\theta D}$, $k_{\theta\theta T}$	$K_{\theta\theta}$ Due to displacement and rotation of balloon relative to steady tension vector at BCP, N-m/rad
$k_{\varphi y}$, $k_{\psi y}$	Tether rolling and yawing moment per unit of y-displacement, Nm/m
$k_{\varphi\psi}$, $k_{\psi\varphi}$	Tether rolling moment per unit of yaw displacement and vice versa, N-m/rad
l	Tether cable length, m
l_{br} or L_{br}	Component of distance from RP to COB, positive for COB ahead of RP, m
l_{cg} or L_{cg}	Component of distance from RP to COM of balloon, positive for COM forward of RP, m
l_{sr} or L_{sr}	Component of distance from RP to COM of balloon structure, positive for COM aft of RP, m
l_{tr} or L_{tr}	Component of distance from RP to BCP, positive for BCP forward of RP, m
L_{PHT}	Distance of CG from MAC of PHT of balloon, m
m_a	Apparent mass of air associated with accelerations of balloon, kg
m_g	Mass of inflation gas, kg
m_s	Balloon structural mass (including bridle, test instr. and payload), kg
m_T	Combined mass of balloon structure and inflation gas, $m_g + m_s$, kg
n	Cable drag per unit length for cable normal to the wind, N/m
S_{ref}, S_{exposed}	Reference area ($\pi D_{\max}{}^2/4$) and exposed planform area of aerostat, m^2
t_{tr} or T_{tr}	Component of distance from RP to BCP, positive for BCP below RP, m
t	Time in seconds
T, T_0, T_1	Tether cable tension, tension at lower and upper ends, respectively, N
u, w	Perturbation velocities of balloon COM along X- and Z-axes, respectively, m/s
V_∞	Steady wind velocity, m/s
V_n	Component of wind velocity normal to cable, $V_\infty \sin\gamma$, m/s
W_s	Structural weight of balloon (including bridle, payload and test instrument), N
w_c	Tether cable weight per unit length, N/m
x_t, z_t	Distance parallel to X- and& Z-axis from RP to COM, +ve for COM forward and below RP, respectively

x_1, z_1	Coordinates of balloon COM with respect to tether cable anchor point, m
α	Perturbation angle of attack, rad
ϵ	Downwash angle, rad
Λ	Tail sweep angle, rad
γ_0, γ_1	Angle between horizontal and tether cable at lower and upper end, respectively, rad
ε	Angle between principal X-axis of balloon and stability axis, rad
η	Real part of characteristic root of stab. equation, damping, s^{-1}
θ	Pitch angle, rad
λ	Characteristic root of stability equations ($\eta \pm i\omega$) and taper ratio
ρ	Atmospheric density, kg/m^3
ω	Imaginary part of characteristic root of stability equations, frequency, rad/s

Subscripts

A, B, C, G	Aerodynamic, buoyancy, tether cable and gravity force terms, respectively.
t	Equilibrium trim condition
0, 1	Lower and upper end of tether cable
$\alpha, \dot{\alpha}$	With respect to α and $\dot{\alpha}\overline{c}/2V_{\infty}$, respectively
q, r	With respect to $q\overline{c}/2\ V_{\infty}$

Abbreviations

BCP, RP	Bridle confluence point and reference point
CG, COB	Center of gravity and center of Buoyancy
COM, SCM	Balloon center of mass and balloon structural center of mass
PHT	Projected horizontal tail
MAC	Mean aerodynamic chord

1 Introduction

The paper presents a systematic approach for longitudinal stability analysis and parametric study on longitudinal stability boundaries of an aerostat tethered from an earth-fixed anchor point and flying in steady wind conditions. Pant et al. [1–3] have reported good amount of research work on sizing, design and fabrication of

Aerostats. Rajani et al. [4, 5] have analyzed dynamic stability of a tethered aerostat. Worth noting work has been reported in the reports available on analytical and experimental determination of stability parameters along with trend study of balloon tethered in wind [6–8]. Authors [9, 10] have analyzed stability along with parametric trend study of a tethered aerostat. The contributions in the area of stability analysis of aerostat [11–14] and tether cable stability and dynamics [15–17] have also been reported earlier. Few references [18–20] have been used for determining some stability parameters. The paper presents mathematical modeling [8] (Sect. 2), estimation of stability characteristics (Sect. 3) and parametric trend study (Sect. 4) showing the effect of various geometrical parameters on longitudinal stability boundaries of a tethered aerostat.

2 Mathematical Modeling

The stability analysis of an aerostat tethered from an earth-fixed anchor point has been carried out under steady wind conditions. The formulations given by Redd et al. [8] have been used for mathematical modeling of the considered aerostat (Fig. 1) tethered in the steady wind conditions.

Figure 2 presents the geometrical parameters and various forces and moments acting on tethered aerostat. The use of theoretical formulations [8] based on considered aerostat configuration was made for the calculation of stability derivatives and analysis.

Figure 3 shows the coordinate system along with forces and moments used for the derivation of equations of motion of the tethered aerostat. Figure 3 also shows tether cable forces at the lower and upper end along with related angles.

Table 1 presents the geometric, mass, inertia and aerodynamic characteristics of the considered aerostat used to carry out the stability analysis. Some dimensional parameters were given, while the others were calculated for the given configuration of the tethered aerostat based on the theoretical formulations [8, 19, 20].

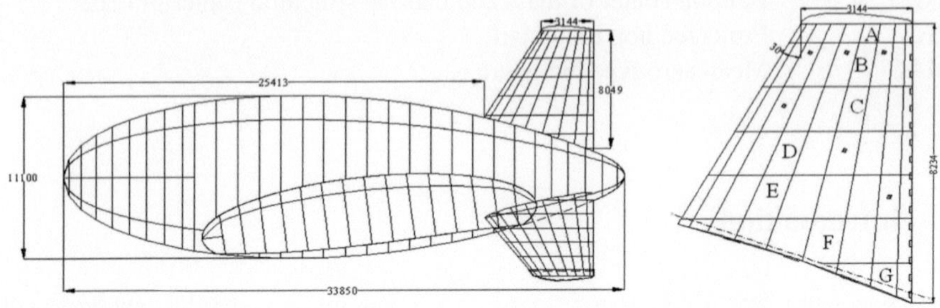

Fig. 1 Dimensions of the aerostat and fin

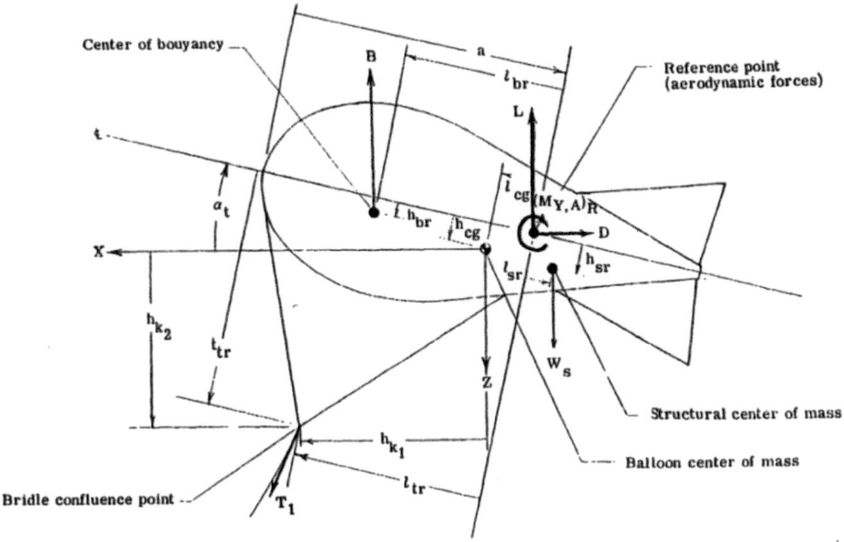

Fig. 2 Geometry of the balloon system [8]

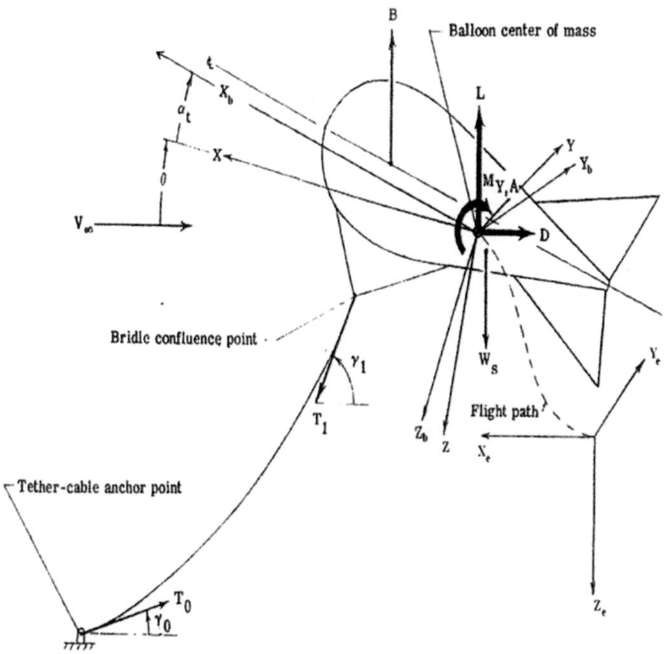

Fig. 3 Coordinate system and forces acting on tethered aerostat [8]

Table 1 Characteristics of the considered aerostat

Parameter (units)	Value	Parameter (units)	Value	Parameter (units)	Value
L_{tr} (m)	5.98	ρ_a (kg-m^{-3})	1.09	bVT (m)	8.1415
T_{tr} (m)	10.9	ρ_{he} (kg-m^{-3})	0.1759	bPVT (m)	5.7572
L_{cg} (m)	−1.92	m_T (kg)	1406	bPHT (m)	11.5145
H_{cg} (m)	0.68	m_{he} (kg)	355.85	SVT (m^2)	44.729
L_{br} (m)	0.31	m_s (kg)	1050.15	SPVT (m^2)	31.63
H_{br} (m)	0.0	$m_{x,a}$ (kg)	488.25	SPHT (m^2)	63.26
L_{sr} (m)	−3.6	$m_{y,a}$ (kg)	2283.6	S_{ref} (m^2)	96.769
H_{sr} (m)	2.4	$m_{z,a}$ (kg)	2283.6	AVT	1.482
l (m)	1000	I_{xx} (kg-m^2)	15,081.44	APVT	1.048
d_c (m)	0.017	I_{yy} (kg-m^2)	150,814.4	APHT	2.096
D_{max} (m)	11.1	I_{zz} (kg-m^2)	150,814.4	LVT (m)	4.2387
L (m)	33.85	B (N)	18,354.51	LPVT (m)	1.4795
c_t (m)	3.144	w_c (N/m)	2.943	LPHT (m)	9.4407
c_r (m)	7.844	C_{Dc}	1.17		
$\bar{c}$ (m)	5.829	λ	0.4		

The motion of the tethered aerostat consists of small perturbations about steady flight reference conditions. A linearized analysis similar to that of a rigid airplane has been used during the mathematical modeling while taking into account the following considerations.

1. The equations of motion are referred to center of mass of the balloon.
2. The balloon is symmetric laterally and has yaw, roll and side slip angles equal to zero in the reference steady-state trimmed condition ($\psi_t, \varphi_t, \beta_t = 0$).
3. The balloon and bridle form a rigid system.
4. The tether cable is flexible, but inextensible and contributes static forces at the bridle confluence point (BCP).
5. The cable weight and drag normal to the cable are needed only for determining the static cable forces, equilibrium shape of the cable and the cable derivatives.

Four different sources of external forces and moments such as aerodynamic, buoyant, tether cable and gravity act on a tethered aerostat. Therefore, the equations of motion of a tethered aerostat can be written as [8].

$$F_{X,A} + F_{X,C} + F_{X,B} + F_{X,G} = m_{x,o}\ddot{x}_e \tag{1a}$$

$$F_{Z,A} + F_{Z,C} + F_{Z,B} + F_{Z,G} = m_{z,o}\ddot{z}_e \tag{1b}$$

$$M_{Y,A} + M_{Y,C} + M_{Y,B} + M_{Y,G} = Iy\ddot{\theta} \tag{1c}$$

The terms $m_{x,o}$, $m_{y,o}$ and $m_{z,o}$ are total aerostat masses in x-, y- and z-directions, respectively, and can be expressed as:

$$m_{x,o} = m_s + m_g + m_{a1} \tag{2a}$$

$$m_{z,o} = m_s + m_g + m_{a3} \tag{2b}$$

The terms m_s, and m_g are the structural mass of aerostat and mass of the gas inside the aerostat. The terms m_{a1} and m_{a3} are apparent masses associated with accelerations in x- and z-directions, respectively. The apparent masses which depend upon the equilibrium trim angle of attack (α_t) are given by the following equations [8].

$$m_{a1} = m_{x,a}\cos^2\alpha_t + m_{z,a}\sin^2\alpha_t, \tag{3a}$$

$$m_{a3} = m_{x,a}\sin^2\alpha_t + m_{z,a}\cos^2\alpha_t \tag{3b}$$

The terms $m_{x,a}$ and $m_{z,a}$ are the apparent masses of the balloon accelerating along the X_b- and Z_b-axes. The mass moments of inertia which depend upon the orientation of the balloon are expressed by the following equations [8].

$$I_x = I_{xx}\cos^2\varepsilon + I_{zz}\sin^2\varepsilon \tag{4a}$$

$$I_y = I_{yy} \tag{4b}$$

$$I_z = I_{xx}\sin^2\varepsilon + I_{zz}\cos^2\varepsilon \tag{4c}$$

$$I_{xz} = 1/2(I_{xx} - I_{zz})\sin^2\varepsilon \tag{4d}$$

The terms I_{xx}, I_{yy} and I_{zz} are the mass moments of inertia about the principal axes, and ε is the angle between the principal X-axis and the stability X-axis. In the present analysis, X_b-, Y_b- and Z_b-axes are considered to be principal axes; hence, $\varepsilon = \alpha_t$.

2.1 Aerodynamic Forces and Moments

The aerodynamic forces and moments at trim conditions in non-dimensional form while neglecting higher order perturbation terms are represented by the following relationships [8].

$$F_{X,A} = -\left[\left(\frac{\rho V_\infty S}{2}\right)(2C_D + C_{D_u})\dot{x}_e\right] - \left[\left(\frac{\rho V_\infty S}{2}\right)(C_{D_\alpha} - C_L)\right]\dot{z}_e - \left[\left(\frac{\rho V_\infty^2 S}{2}\right)(C_{D_\alpha} - C_L)\right]\theta - \left(\frac{\rho V_\infty^2 S}{2}\right)C_D \tag{5a}$$

$$F_{Z,A} = -\left[\left(\frac{\rho V_\infty S}{2}\right)(2C_L + C_{L_u})\right]\dot{x}_e - \left[\left(\frac{\rho S\overline{c}}{4}\right)C_{L_{\dot{\alpha}}}\right]\ddot{z}_e - \left[\left(\frac{\rho V_\infty S}{2}\right)(C_{L_\alpha} + C_D)\right]\dot{z}_e - \left[\frac{\rho V_\infty S\overline{c}}{4}\left(C_{L_{\dot{\alpha}}} + C_{L_q}\right)\right]\dot{\theta} - \left[\frac{\rho V_\infty^2 S}{2}(C_{L_\alpha} + C_D)\right]\theta + \frac{\rho V_\infty^2 S}{2}C_L \tag{5b}$$

$$M_{Y,A} = \left[\frac{\rho V_\infty S\overline{c}}{2}(2C_m + C_{m_u})\right]\dot{x}_e + \left[\frac{\rho S(\overline{c})^2}{4}C_{m_{\dot{\alpha}}}\right]\ddot{z}_e + \left(\frac{\rho V_\infty S\overline{c}}{2}C_{m_\alpha}\right)\dot{z}_e + \frac{\rho V_\infty S(\overline{c})^2}{4}(C_{m_{\dot{\alpha}}} + C_{m_q})\dot{\theta} + \left(\frac{\rho V_\infty^2 S\overline{c}}{2}C_{m_\alpha}\right)\theta + \frac{\rho V_\infty^2 S\overline{c}}{2}C_m \tag{5c}$$

2.2 Tether Cable Forces and Moments

The tether cable forces and moments are expressed as [16]:

$$F_{X,C} = -k_{xx}x_e - k_{xz}z_e - (k_{x\theta} + T_1 \sin\gamma_1)\theta + T_1\cos\gamma_1 \tag{6a}$$

$$F_{Z,C} = -k_{zx}x_e - k_{zz}z_e + \left(T_1\cos\gamma_1 + k_{y\theta}\right)\theta + T_1\sin\gamma_1 \tag{6b}$$

$$M_{Y,C} = -k_{\theta x}x_e - k_{\theta z}z_e - k_{\theta\theta}\theta - h_{k_1}T_1\sin\gamma_1 + h_{k_2}T_1\cos\gamma_1 \tag{6c}$$

where

$$h_{k_1} = \left(\mathrm{l}_{tr} - \mathrm{l}_{cg}\right)\cos\alpha_t + \left(t_{tr} - h_{cg}\right)\sin\alpha_t$$

$$h_{k_2} = \left(t_{tr} - h_{cg}\right)\cos\alpha_t - \left(\mathrm{l}_{tr} - \mathrm{l}_{cg}\right)\sin\alpha_t,$$

$$k_{x\theta} = h_{k_2}k_{xx} - h_{k_1}k_{x2}, \qquad k_{z\theta} = h_{k_2}k_{zx} - h_{k_1}k_{zz}$$

$$k_{\theta x} = h_{k_2}k_{xx} - h_{k_1}k_{zx}, \qquad \mathrm{k}_{\theta z} = h_{k_2}k_{xz} - h_{k_1}k_{zz}$$

$$k_{\theta\theta} = k_{\theta\theta_D} + k_{\theta\theta_T}$$

$$k_{\theta\theta_D} = h_{k_2}^2 k_{xx} - h_{k_2} h_{k_1} (k_{xz} + k_{zx}) + h_{k_1}^2 k_{zz}$$

$$k_{\theta\theta_T} = h_{k_2} (T_1 \sin \gamma_1) + h_{k_1} (T_1 \cos \gamma_1)$$

$$k_{y\varphi} = -h_{k_2} k_{yy}, \quad k_{y\psi} = h_{k_1} k_{yy}, \quad k_{\varphi y} = k_{y\varphi}$$

$$k_{\varphi\varphi} = h_{k_2}^2 k_{yy}, \qquad k_{\psi\psi} = h_{k_1}^2 k_{yy}$$

$$k_{\varphi\psi} = -h_{k_1} h_{k_2} k_{yy}, \quad k_{\psi y} = k_{y\psi}, \quad k_{\psi\varphi} = k_{\varphi\psi}$$

2.3 Buoyancy Forces and Moments

The expressions for the buoyancy forces and moments about the center of mass in the stability axis system can be expressed assuming small perturbation angles as [16]:

$$F_{X,B} = B\theta \tag{7a}$$

$$F_{Z,B} = -B \tag{7b}$$

$$\begin{aligned} M_{Y,B} = {} & B\left[\left(\mathrm{l}_{br} - \mathrm{l}_{cg}\right)\cos\alpha_t - \left(h_{cg} - h_{br}\right)\sin\alpha_t\right] \\ & - B\left[\left(h_{cg} - h_{br}\right)\cos\alpha_t + \left(\mathrm{l}_{br} - \mathrm{l}_{cg}\right)\sin\alpha_t\right]\theta \end{aligned} \tag{7c}$$

2.4 Gravity Forces and Moments

The component due to structural weight of balloon is considered during the formulation of equations of motion for gravity forces. The effects of apparent mass and lifting gas are already included in the coefficients of the acceleration and buoyancy terms, respectively. The forces and moments due to gravity for small perturbation angles are determined by [8]:

$$F_{X,G} = -W_s\theta \tag{8a}$$

$$F_{Z,G} = -W_s \tag{8b}$$

$$M_{Y,G} = W_S\left[\left(l_{sr} + l_{cg}\right)\cos\alpha_t - \left(h_{sr} - h_{cg}\right)\sin\alpha_t\right] - W_S\left[\left(h_{sr} - h_{cg}\right)\cos\alpha_t + \left(l_{sr} + l_{cg}\right)\sin\alpha_t\right]\theta \tag{8c}$$

3 Estimation of the Stability Characteristics

After the mathematical modeling, the stability characteristics (roots/eigen values) of the considered aerostat can be estimated by executing the following steps:

1. Calculate the trim angle of attack.
2. Obtain the aerodynamic parameters dependent on trim angle of attack for the steady-state trim condition.
3. Calculate the value of tensions in the cable at the upper and lower ends.
4. Use the value of tensions to obtain tether cable derivatives.
5. Obtain the stability equations by putting the equilibrium part of the balloon's equations of motion to zero.
6. Convert the above stability equations in the matrix form and obtain the roots/eigen values by using the results obtained in the steps 1 to 4.

3.1 Balloon Equations of Motion

After combining all the expressions for each of the external forces and moments (such as aerodynamic, buoyancy, cable-tether and gravity), the following resulting equations of motion (16) about the balloon COM can be obtained.

X-Force

$$m_x\ddot{x}_e + \left[\frac{\rho V_\infty S}{2}\left(2C_D + C_{D_u}\right)\right]\dot{x}_e + k_{xx}x_e + \left[\frac{\rho V_\infty S}{2}\left(C_{D_\alpha} - C_L\right)\right]\dot{z}_e + k_{xz}z_e + \left[k_{x\theta} + \frac{\rho V_\infty^2 S}{2}\left(C_{D_\alpha} - C_L\right) - (B - W_s) + T_1\sin\gamma_1\right]\theta + \frac{\rho V_\infty^2 S}{2}C_D - T_1\cos\gamma_1 = 0 \tag{9a}$$

Z-Force

$$m_z\ddot{z}_e + \frac{\rho V_\infty S}{2}(2C_L + C_{L_u})\dot{x}_e + k_{zx}x_e + \frac{\rho V_\infty S}{2}(C_{L_\alpha} + C_D)\dot{z}_e + k_{zz}z_e + \frac{\rho V_\infty S\overline{c}}{4}(C_{L_{\dot{\alpha}}} + C_{L_q})\dot{\theta} + \left(k_{z\theta} + \frac{\rho V_\infty^2 S}{2}(C_{L_\alpha} + C_D) - T_1\cos\gamma_1\right)\theta$$

$$+\frac{\rho V_\infty^2 S}{2} C_L + B + W_S - T_1 \sin\gamma_1 = 0 \tag{9b}$$

Pitching Moment

$$\begin{aligned}
&-\left[\frac{\rho V_\infty S\overline{c}}{2}\left(2C_m + C_{m_u}\right)\right]\dot{x}_e + k_{\theta x}x_e - \left[\frac{\rho S\overline{c}^2}{4}C_{m_{\dot\alpha}}\right]\ddot{z}_e \\
&-\left(\frac{\rho V_\infty S\overline{c}}{2}C_{m_\alpha}\right)\dot{z}_e + k_{\theta z}z_e + I_y\ddot{\theta} - \left[\frac{\rho V_\infty S\overline{c}^2}{4}\left(C_{m_{\dot\alpha}} + C_{m_q}\right)\right]\dot{\theta} \\
&+\left(k_{\theta\theta} + M_{s_1} - \frac{\rho V_\infty^2 S\overline{c}}{2}C_{m_\alpha}\right)\theta - \frac{\rho V_\infty^2 S\overline{c}}{2}C_m + h_{k_1}T_1\sin\gamma_1 \\
&- h_{k_2}T_1\cos\gamma_1 - M_{s_2} = 0
\end{aligned} \tag{9c}$$

$$\begin{aligned}
M_{s_1} &= \left[\left(\mathrm{l}_{br} - \mathrm{l}_{cg}\right)B + \left(\mathrm{l}_{sr} + \mathrm{l}_{cg}\right)W_s\right]\sin\alpha_t \\
&\quad + \left[\left(h_{cg} - h_{br}\right)B + \left(h_{sr} - h_{cg}\right)W_s\right]\cos\alpha_t \\
M_{s_2} &= \left[\left(\mathrm{l}_{br} - \mathrm{l}_{cg}\right)B + \left(\mathrm{l}_{sr} + \mathrm{l}_{cg}\right)W_s\right]\cos\alpha_t \\
&\quad - \left[\left(h_{cg} - h_{br}\right)B + \left(h_{sr} - h_{cg}\right)W_s\right]\sin\alpha_t \\
m_x &= m_{x,o} \quad \text{and} \quad m_z = m_{z,o} + \frac{\rho S\overline{c}}{4}C_{L_{\dot\alpha}}
\end{aligned}$$

3.2 *Equilibrium Trim Conditions*

In the mathematical model used for calculating the stability characteristic, it is seen that all the aerodynamic parameters are dependent on the angle of attack and it is required to calculate the angle of attack at which the steady-state trimmed condition for the balloon is achieved, this angle of attack is called the trim angle of attack. The steady-state trimmed conditions can be obtained by setting the perturbation quantities of Eq. (9a–9c) equal to zero.

$$\frac{\rho V_\infty^2 S}{2} C_D - T_1 \cos\gamma_1 = 0 \tag{10a}$$

$$\frac{\rho V_\infty^2 S}{2} C_L + B - W_s - T_1 \sin\gamma_1 = 0 \tag{10b}$$

$$-\frac{\rho V_\infty^2 S\tau}{2} C_m + h_{k_1}T_1\sin\gamma_1 - h_{k_2}T_1\cos\gamma_1 - M_{s_2} = 0 \tag{10c}$$

Substitute Eq. (10a–10b) into Eq. (10c) to eliminate the cable tension and angle to obtain the following trim equation:

$$h_{k_1}\left(\frac{\rho V_\infty^2 S}{2} C_L + B - W_s\right) - h_{k_2}\left(\frac{\rho V_\infty^2 S}{2} C_D\right) - \frac{\rho V_\infty^2 S\tau}{2} C_m - M_{s_2} = 0 \quad (11)$$

Equation (11) can be solved by Newton iterations to find the equilibrium trim angle of attack (α_t) for various wind velocities, provided the aerodynamic coefficients C_L, C_D and C_m are known functions. The calculated α_t can be used to solve the Eq. (10a–10c) to find and followed by the evaluation of α-dependent stability coefficients.

3.3 Formulations for Calculation of Stability Derivatives

The expressions for the longitudinal stability coefficient/derivatives calculated in the previous step are based on the theoretical formulation corresponding to CG location. The derivative based on the aerostat configuration has been calculated for projected horizontal (PHT). Lift curve slope expression given in Eq. (12) uses the values of constants of the respective tail (PHT).

$$C_{L_{\alpha_t}} = \frac{(2\pi A)}{\left(2 + \sqrt{4 + \frac{A^2\beta^2}{\eta^2}\left(1 + \frac{\tan^2\Lambda}{\beta^2}\right)}\right)} * \frac{S_{\text{exposed}}}{S_{\text{ref}}} \quad (12)$$

where $C_{L_{\alpha_t}}$ is the lift curve slope of the tail.

Longitudinal Derivatives (PHT).

$$C_L = 0.0061 + 1.2\alpha + C_{L_{\alpha_t}}\alpha + \eta C_{D_c}\frac{S_P}{S_{\text{ref}}}\alpha^2$$

$$C_{L_\alpha} = 1.2 + C_{L_{\alpha_t}} + 2\eta C_{D_c}\frac{S_P}{S_{\text{ref}}}\alpha$$

$$C_{L_{\dot\alpha}} = C_{L_q}\frac{\mathrm{d}\in}{\mathrm{d}\alpha}$$

$$C_{L_q} = 2C_{L_{\alpha_t}}\frac{L_{\text{PHT}}}{D}$$

$$C_D = 0.0396 + \frac{C_L^2}{\pi e \text{A}}$$

$$C_{D_\alpha} = 2\frac{C_L}{\pi e A}C_{L_\alpha}$$

$$C_m = -0.02 + 0.04832\alpha + \eta C_{L_{\alpha_t}}\left(1 - \frac{\mathrm{d}e}{\mathrm{d}\alpha}\right)\frac{L_{\mathrm{PHT}}}{D}\alpha$$

$$C_{m_\alpha} = 0.048326 + \eta C_{L_{\alpha_t}}\left(1 - \frac{\mathrm{d}e}{\mathrm{d}\alpha}\right)\frac{L_{\mathrm{PHT}}}{D}\alpha$$

$$C_{m_q} = -2C_{L_{\alpha_t}}\left(\frac{L_{\mathrm{PHT}}}{D}\right)^2$$

$$C_{m_{\dot\alpha}} = C_{m_q}\tau\frac{\mathrm{d}\in}{\mathrm{d}\alpha}$$

where $\tau = \left(\frac{V_t}{V}\right)^2$.

3.4 Equilibrium Cable Shape

The forces acting on tether cable of length, l (Fig. 4) are the tension, cable weight and drag normal to the cable. Drag along the cable has been neglected. The normal drag force per unit length depends on the component of wind velocity normal to the cable V_n, the drag cable coefficient C_{D_c} and cable diameter $\mathrm{d_c}$ and can be expressed as [8]:

$$n = C_{D_c} d_c \frac{1}{2}\rho V_n^2 \tag{13}$$

Tension (T_1) at upper end of the cable using tension $\left[\frac{\mathrm{d}T}{T} = -\frac{\overline{p}}{\mathrm{q}}\left(\frac{\mathrm{d}f}{\overline{q}+\overline{p}-f} + \frac{\mathrm{d}f}{\overline{q}-\overline{p}+f}\right)\right]$ is given by

$$T_1 = T_{\tau 1/\tau} \tag{14}$$

where $\tau(\gamma) = \left(\frac{\overline{q}+\overline{p}-\cos\gamma}{\overline{q}-\overline{p}+\cos\gamma}\right)^{\frac{\overline{p}}{\overline{q}}}$, $\overline{p} = \frac{w_c}{2n}$, $\overline{q} = \sqrt{1+(\overline{p})^2}$, $f = \cos\gamma$.

For the known parameters such as l $\left(\mathrm{d}l = \left(\frac{T_1}{n\tau_1}\right)\frac{\tau}{(\sin^2\gamma+2\overline{p}\cos\gamma)}\mathrm{d}\gamma\right)$, n, w_c, T_1 and γ_1, the following expressions can be used to determine the coordinates T_1 and γ_1 at upper end and T_0 and γ_0 at the lower end.

$$\overline{\lambda}_0 = \overline{\lambda}_1 - \frac{n\tau_1 l}{T_1} \quad \text{and} \quad T_0 = T_{1\tau 0/\tau 1} \tag{15}$$

$$\tilde{x}_1 = \frac{T_1}{n\tau_1}\int_{\gamma_0}^{\gamma_1}\frac{\tau\cos\gamma}{(\sin^2\gamma + 2\overline{p}\cos\gamma)}\mathrm{d}\gamma \quad \text{where} \quad \mathrm{d}\sigma = \frac{\tau\cos\gamma}{(\sin^2\gamma + 2\overline{p}\cos\gamma)}\mathrm{d}\gamma \tag{16}$$

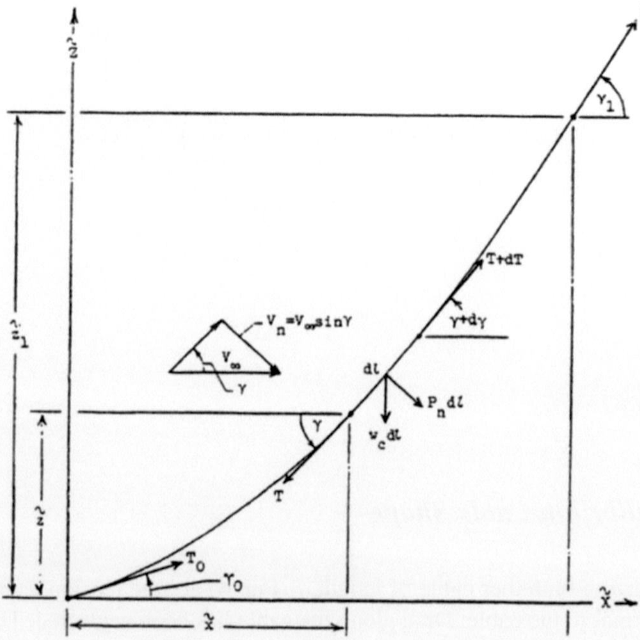

Fig. 4 Forces acting on the tether cable [8]

$$\tilde{z}_1 = \frac{T_1 - T_0}{w_c} \quad \text{where} \quad d\tilde{z} = dl \sin\gamma = \frac{dT}{w_c} \tag{17}$$

where $\overline{\lambda}(\gamma) = \int\limits_0^{\gamma} \frac{\tau(\gamma)}{(\sin^2\gamma + 2\overline{p}\cos\gamma)} d\gamma$ $\overline{\lambda}_0 = \overline{\lambda}(\gamma_0)$ and $\overline{\lambda}_1 = \overline{\lambda}(\gamma_1)$.

3.5 Cable Force Derivatives

Consider cable in its equilibrium position. If upper end is slowly displaced in the $\tilde{x}\tilde{z}$—plane from its original position $\tilde{x}_1, \tilde{z}_1$ to a new position the resultant x- and z-force increments are

$$dF_x = k_{xx} d\tilde{x} + k_{xz} d\tilde{z} \tag{18a}$$

$$dF_z = k_{zx} d\tilde{x} + k_{zz} d\tilde{z} \tag{18b}$$

The cable derivatives (spring constants) k_{xx}, k_{xz}, k_{zx} and k_{zz} for the longitudinal case can be expressed as [8]:

$$k_{xx} = \frac{1}{\delta}\left[T_1 \cos\gamma_1(\sin\gamma_1 - \sin\gamma_0) + n(z_1 - 1\sin\gamma_0)\sin^3\gamma_1\right] \quad (19a)$$

$$k_{xz} = \frac{1}{\delta}\left[T_1 \cos\gamma_1(\cos\gamma_0 - \cos\gamma_1) + n(1\cos\gamma_0 - \tilde{x}_1)\sin^3\gamma_1\right] \quad (19b)$$

$$k_{zx} = \frac{1}{\delta}\left[T_1 \sin\gamma_1(\sin\gamma_1 - \sin\gamma_0) - \left(w_c + n\sin^2\gamma_1\cos\gamma_1\right)(\tilde{z}_1 - 1\sin\gamma_0\right] \quad (19c)$$

$$k_{zz} = \frac{1}{\delta}\left[T_1 \sin\gamma_1(\cos\gamma_0 - \cos\gamma_1) - \left(w_c + n\sin^2\gamma_1\cos\gamma_1\right)(1\cos\gamma_0 - \tilde{x}_1)\right] \quad (19d)$$

where $\delta = x_1(\sin\gamma_1 - \sin\gamma_0) + z_1(\cos\gamma_0 - \cos\gamma_1) - 1\sin(\gamma_1 - \gamma_0)$.

The single lateral cable derivative determined by considering a small force dF_Y to act in the y-direction on the upper end of the cable is given by the following expression.

$$dF_Y = k_{yy}dy \quad (20)$$

where $k_{yy} = \frac{n\sqrt{\tau_1\left(\sin^2\gamma_1 + 2\overline{p}\cos\gamma_1\right)}}{\int_{\gamma_0}^{\gamma_1}\sqrt{\frac{\tau(\gamma)}{\left(\sin^2\gamma + 2\overline{p}\cos\gamma\right)}}d\gamma}$.

3.6 Stability Equations (Longitudinal)

The stability equations are obtained by setting the equilibrium trim portions of the equations of motion (Eq. 10a–10f) equal to zero. The following working forms of the stability equations [3] written about the balloon center of mass are obtained.

***X*-Force.**

$$m_x\ddot{x}_e + \left[\frac{\rho V_\infty S}{2}\left(2C_D + C_{D_u}\right)\right]\dot{x}_e + k_{xx}x_e + \left[\frac{\rho V_\infty S}{2}\left(C_{D_\alpha} - C_L\right)\right]\dot{z}_e$$
$$+ k_{xz}z_e + \left[k_{x\theta} + \frac{\rho V_\infty^2 S C_{D_\alpha}}{2}\right]\theta = 0 \quad (21a)$$

Z-Force.

$$m_z\ddot{z}_e + \frac{\rho V_\infty S}{2}(2C_L + C_{L_u})\dot{x}_e + k_{zx}x_e + \frac{\rho V_\infty S}{2}(C_{L_\alpha} + C_D)\dot{z}_e$$
$$+ k_{zz}z_e + \frac{\rho V_\infty S\overline{c}}{4}(C_{L_{\dot{\alpha}}} + C_{L_q})\dot{\theta} + \left(k_{z\theta} + \frac{\rho V_\infty^2 S C_{L_\alpha}}{2}\right)\theta = 0 \quad (21b)$$

Pitching Moment.

$$-\left[\frac{\rho V_\infty S\overline{c}}{2}\left(2C_m + C_{m_u}\right)\right]\dot{x}_e + k_{\theta x}x_e - \left[\frac{\rho S\overline{c}^2}{4}C_{m_{\dot{\alpha}}}\right]\ddot{z}_e - \left(\frac{\rho V_\infty S\overline{c}}{2}C_{m_\alpha}\right)\dot{z}_e + k_{\theta z}z_e$$
$$+ I_y\ddot{\theta} - \left[\frac{\rho V_\infty S\overline{c}^2}{4}\left(C_{m_{\dot{\alpha}}} + C_{m_q}\right)\right]\dot{\theta}$$
$$+ \left(k_{\theta\theta} + M_{s_1} - \frac{\rho V_\infty^2 S\overline{c}}{2}C_{m_\alpha}\right)\theta = 0 \quad (21c)$$

Using the mathematical model, the stability equations can be written in the state space form as given below:

$$\frac{\mathrm{d}x}{\mathrm{d}t} = Ax + Bu \quad (22)$$

where A is the characteristic matrix and B is the input matrix.

Since no control input is being used, the matrix A gives the characteristics of the aerostat system. The equation for longitudinal and lateral stability case can be expressed in the following matrix form, respectively.

$$\begin{bmatrix} \dot{u} \\ \dot{w} \\ \dot{q} \\ \dot{\theta} \\ \dot{x} \\ \dot{z} \end{bmatrix} = A \begin{bmatrix} u \\ w \\ q \\ \theta \\ x \\ z \end{bmatrix} \quad (23)$$

The roots of characteristic equation obtained by computing stability matrix A for longitudinal and lateral case give an insight into the stability of the system.

4 Effect of Geometrical Parameters on Longitudinal Stability Boundaries

The computed values of longitudinal frequencies (ω) and damping rates (η) for the considered aerostat have been plotted as a function of wind velocity in Fig. 5a, b and in root locus form in Fig. 5c. Figure 5a, b indicates that the considered aerostat has three oscillatory modes of motion for the given range of the wind velocities. It can be observed from Fig. 5b that the aerostat was longitudinally stable except below wind velocity of 2 m/s at which one of the roots becomes positive. This fact is also evident from the negative slope of the plot between pitching moment coefficient and angle of attack as shown in Fig. 5d. It could also be observed from Fig. 5b that mode 2 splits into two real non-oscillatory modes above wind velocity of 19 m/s and again merged into one at 35 m/s.

Fig. 5 **a** Variation of ω with V for longitudinal case. **b** Variation of η with V for longitudinal case. **c** ω versus η (Root locus plot for longitudinal case). **d** C_m versus α for longitudinal case

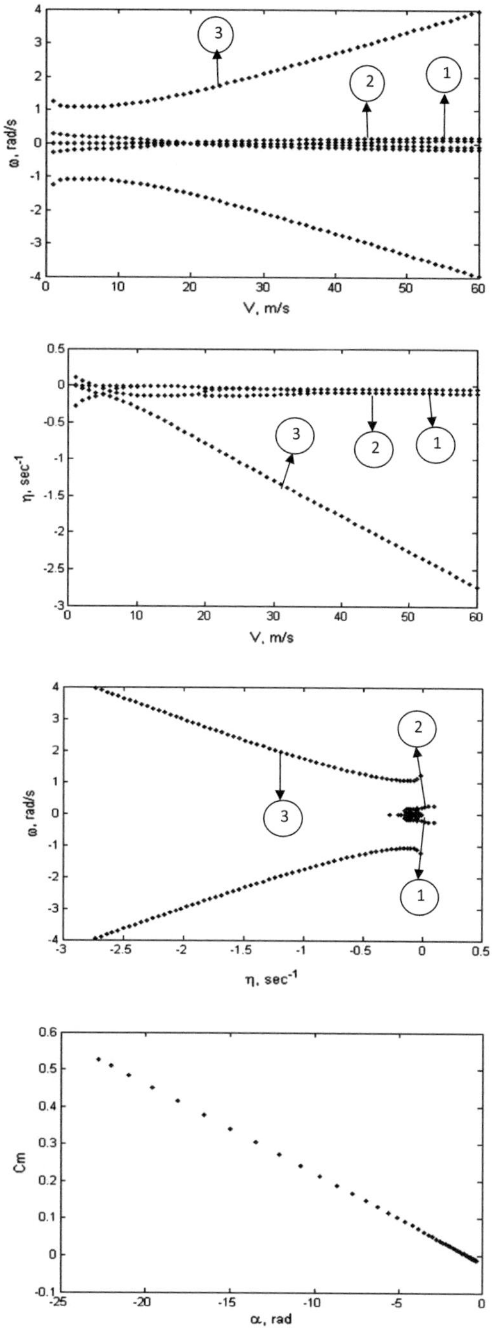

Next, geometrical parameters were varied to see the effect on longitudinal stability boundaries of the considered aerostat. The results showing the effect of different parameters on the stability boundaries for a range of speed have been presented in the graphical form. Figures 6, 7, 8, 9, 10, 11, 12 and 13 show that the aerostat is unstable below the speed of 2 m/sec and in the region bounded by the two curved/straight boundaries. The unstable region increases or decreases with increase or decrease in the values of most of the dimensional parameters of the considered aerostat. Very little or negligible effect on stability boundaries was observed for some parameters.

It can be observed from Figs. 6, 7, 8, 9, 10, 11, 12 and 13 that the parameters such as L_{tr}, T_{tr}, L_{br}, L_{sr}, C.G. (moment arm), l, d_c, w_c affect the stability boundaries strongly while the parameters such as L_{cg}, H_{cg}, H_{br} and H_{sr} have very little or negligible effect on the stability characteristics/boundaries of the aerostat. It can be observed that the decrease in L_{tr} (the horizontal component of distance between RP and BCP) decreases the unstable region while decrease in T_{tr} (the vertical component of distance between RP and BCP) increases the unstable region (Fig. 6a, b). The change in horizontal (L_{cg}) or vertical (H_{cg}) component of distance from RP to COM has very little or negligible effect on the stability boundaries (Fig. 7a, b). Increase in the value of horizontal component of distance from RP to COB (L_{br}) and COM

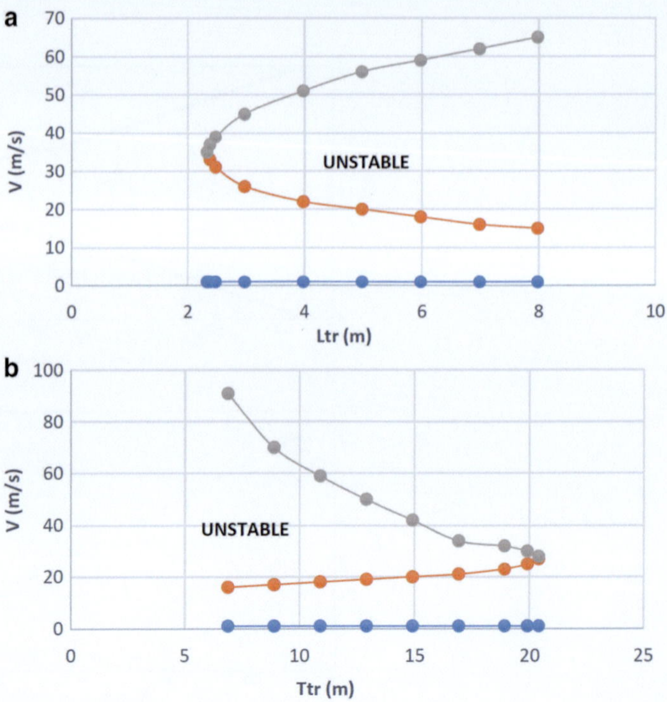

Fig. 6 **a** Effect of L_{tr} on longitudinal stability boundary **b** Effect of T_{tr} on longitudinal stability boundary

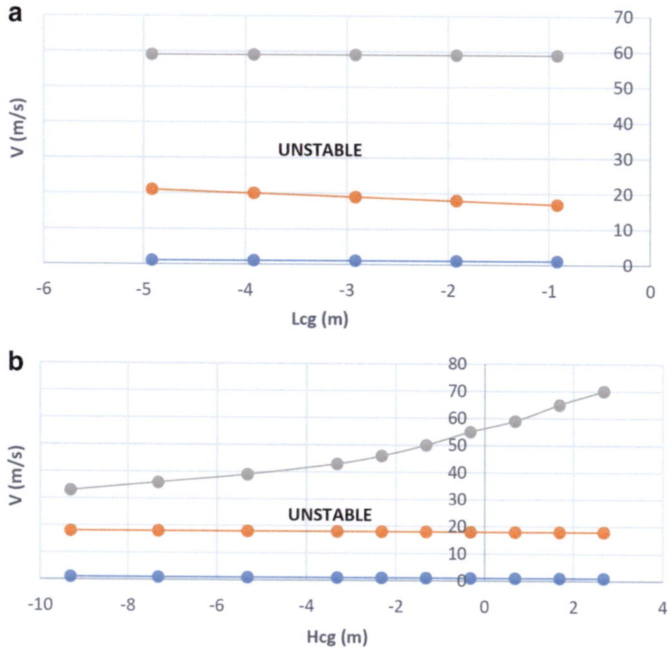

Fig. 7 **a** Effect of L_{cg} on longitudinal stability boundary **b** Effect of H_{cg} on longitudinal stability boundary

of structure (L_{sr}) decreases the unstable regions while the vertical components (H_{br} and H_{sr}) have negligible effect (Figs. 8a, b and 9a, b).

Reduction in tether cable length (l), cable diameter (d_c) and cable weight (w_c) leads to the reduction in the unstable region (Figs. 10, 11 and 12). Increase in the horizontal tail moment arm reduces the unstable region (Fig. 13).

5 Conclusion

Longitudinal stability analysis and effect of variation of geometrical parameters on longitudinal stability boundaries for a balloon tethered in a steady wind has been presented. Equations of motion of the considered aerostat included aerodynamic, tether cable, buoyancy and gravity forces along with aerodynamic apparent mass and structural mass terms. After mathematical modeling, the roots of the characteristic stability equation were computed and plotted for various steady-wind conditions. It was observed from graphical presentations that the considered aerostat was stable longitudinally. Later on, parametric trend study was carried out to show the influence of various dimensional and aerodynamic parameters of aerostat on longitudinal stability boundaries for a wide range of steady-wind speeds. The study suggests that

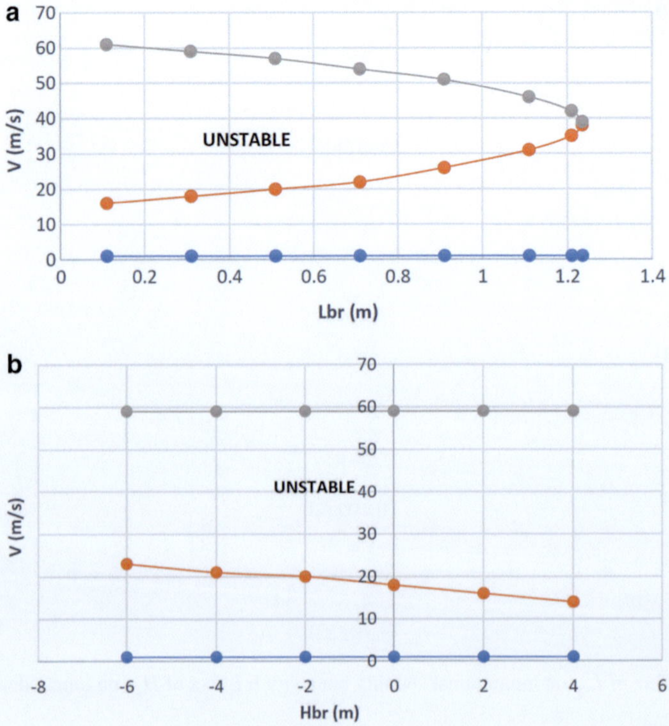

Fig. 8 **a** Effect of L_{br} on longitudinal stability boundary **b** Effect of H_{br} on longitudinal stability boundary

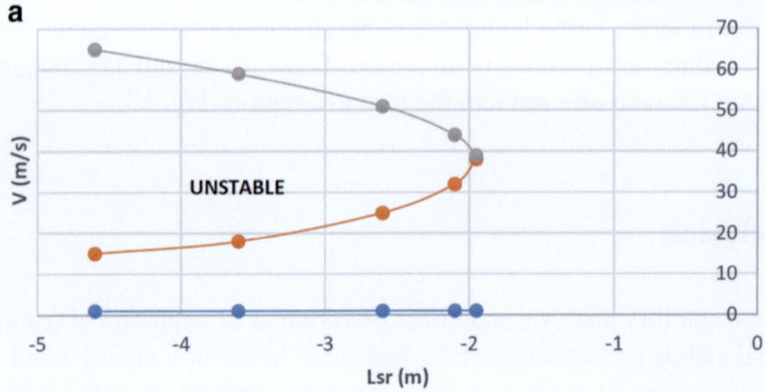

Fig. 9 **a** Effect of L_{sr} on longitudinal stability boundary **b** Effect of H_{sr} on longitudinal stability boundary

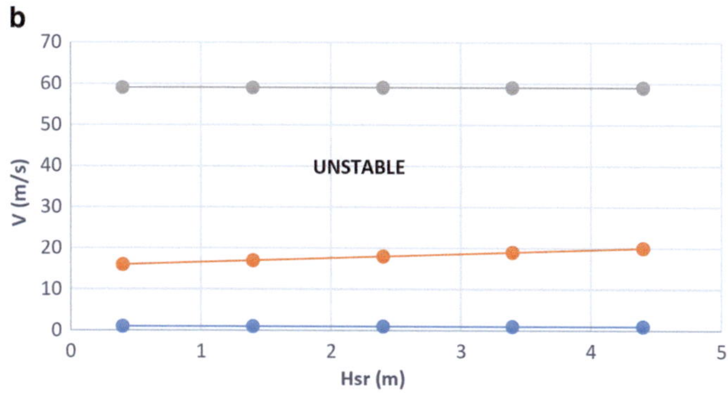

Fig. 9 (continued)

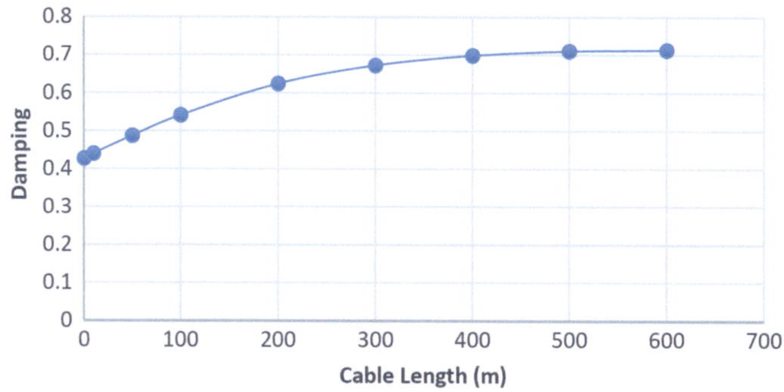

Fig. 10 Effect of cable length (m) on longitudinal stability

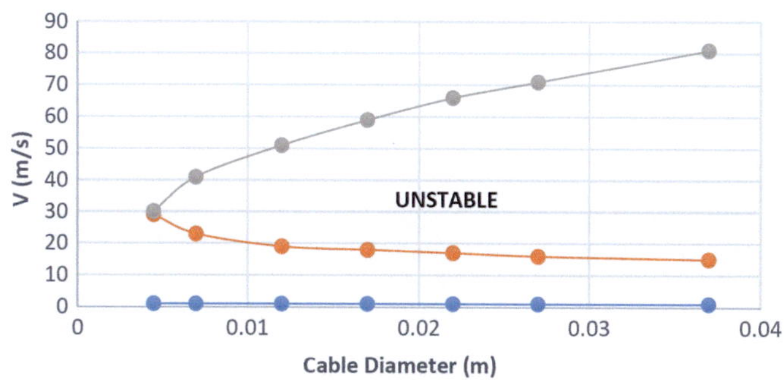

Fig. 11 Effect of cable diameter (d_c) on longitudinal stability boundary

the judicious and feasible choice of various geometrical parameters can be utilized to design a new tethered aerostat which can remain stable for a wide range of wind speeds. The limitation of the stability analysis carried out was that the downwash has been neglected and provides the basis for the future scope.

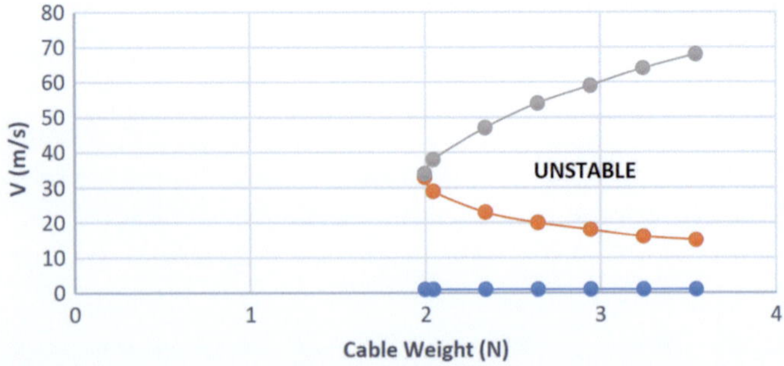

Fig. 12 Effect of cable weight (w_c) on longitudinal stability boundary

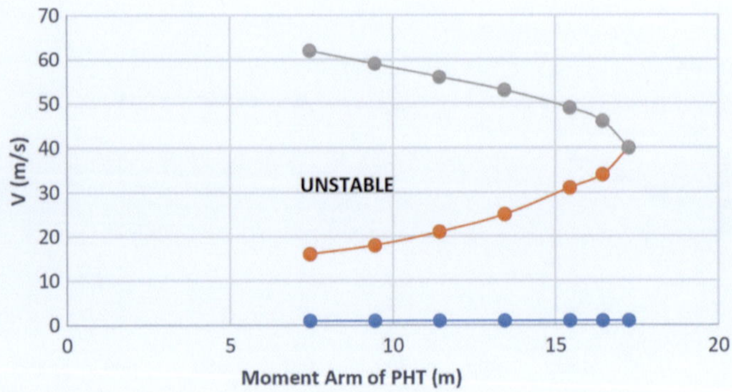

Fig. 13 Effect of moment arm (PHT) on longitudinal stability boundary

References

1. Gupta P, Pant RS (Dec 2005) A methodology for initial sizing and conceptual design studies of an aerostat, international seminar on challenges on aviation technology, integration and operations (CATIO-05), technical. Sessions of 57th annual general meeting of aeronautical society of India
2. Gawande VN, Bilaye P, Gawale AC, Pant RS, Desai UB (Sept 2007) Design and fabrication of an aerostat for wireless communication in remote areas. System technology conference, Belfast, Northern Ireland, UK
3. Raina AA, Bhandari KM, Pant RS (6–10 April 2009) Conceptual design of a high altitude aerostat for study of snow patterns. Proceedings of international symposium on snow and avalanches (ISSA-09), SASE, Manali, India
4. Rajani A, Pant RS, Sudhakar K (4–7 May 2009) Dynamic stability analysis of a tethered aerostat. Proceedings of 18th AIAA lighter-than-air system technology conference, seattle, Washington, USA
5. Rajani A, Pant RS, Sudhakar K (Sept–Oct 2010) Dynamic stability analysis of a tethered aerostat. J Aircraft 47(5)
6. Redd LT, Benett RM, Bland SR (Sept 1972) Analytical and experimental investigation of stability parameters for a balloon tethered in wind, 7th AF cambridge research laboratories scientific balloon symposium. Portsmouth, N.H.

7. Redd LT, Benett RM, Bland SR (1973) Experimental and analytical determination of stability parameters for a balloon tethered in wind, numeric value TD D-2021
8. Redd, L.T., Benett, R.M. and Bland, S.R., "Stability Analysis and Trend Study of a Balloon Tethered in Wind, with Comparisons", NASA TN D-7272, October, 1973.
9. Srivastava S (April 2009) Stability analysis and parameter trend study of single tether aerostats. M. Tech thesis, DAE, IIT, Kanpur
10. Rakesh K et al (2011) Parametric trend study during stability analysis of a tethered aerostat. J Aerospace Sci Technol 63(2)
11. Khouri GA, Gillett JD (1999) Airship technology. Cambridge University Press, UK
12. Delaurier JD (1972) A stability analysis for tethered aerodynamically shaped balloons. J Aircr 9(9):646–651
13. Lambert C, Nahon M (July–Aug 2003) Stability analysis of a tethered aerostat. J Aircraft 40(4)
14. Li Y, Nahon M (Nov–Dec 2007) Modeling and simulation of airship dynamics. J Guidance, Control Dyn 30(6)
15. Neumark S (1963) Equilibrium configurations of flying cables of captive balloons and cable derivatives for stability calculations, R & M no. 3333, Brit., A.R.C.
16. Delaurier JD (Dec 1970) A first order theory for predicting the stability of cable towed and tethered bodies where the cable has a general curvature and tension variation, VKI-TN-68, Von Karman Institute of Fluid Dynamics
17. Delaurier JD (1972) A stability analysis of cable-body system totally immersed in fluid stream. Numeric Value CR-2021
18. Raymer DP (2006) Aircraft design: a conceptual approach, 4th edn. AIAA Education Series, New York, NY
19. Etkin B, Lloyd DR (1996) Dynamics of flight: stability and control, 3rd edn. Wiley
20. Nelson RC. Flight stability and automatic control, 2nd edn. McGraw-Hill, 98

CPACS LTA—Using Common Data Structures for Visualization and Optimization of Airship Designs

Carl Nenoff, Alexander Richter, and David Schlipf

1 Introduction

In this section, we present the background and objectives of our work.

1.1 Background

Powered near equilibrium aerostats, commonly referred to as airships, could and should be a solution of future transport problems. They are eco-friendly with low demand to infrastructure and high safety. Reassessing past designs of successfully operated airships and optimizing preliminary designs are foundations of LTA development.

Optimizing early designs has a huge impact on the overall cost of a project and is at the same time cheap compared to changes made in later design phases. MDO is a technique originated in the aerospace industry. Tools and data commonly used in aircraft industries are not yet implemented for the use with airships. Figure 1 provides an overview of different aircraft categories.

C. Nenoff (✉) · D. Schlipf
Flensburg University of Applied Sciences, Kanzleistraße 91-93, 24943 Flensburg, Germany
e-mail: carl@aerarium.de

D. Schlipf
e-mail: david.schlipf@hs-flensburg.de

A. Richter
Technical University Berlin, Marchstraße 12-14, 10587 Berlin, Germany
e-mail: alexander.richter@bership.de

D. Shukla (ed.), *Lighter Than Air Systems*, Lecture Notes in Mechanical Engineering,
https://doi.org/10.1007/978-981-19-6049-9_2

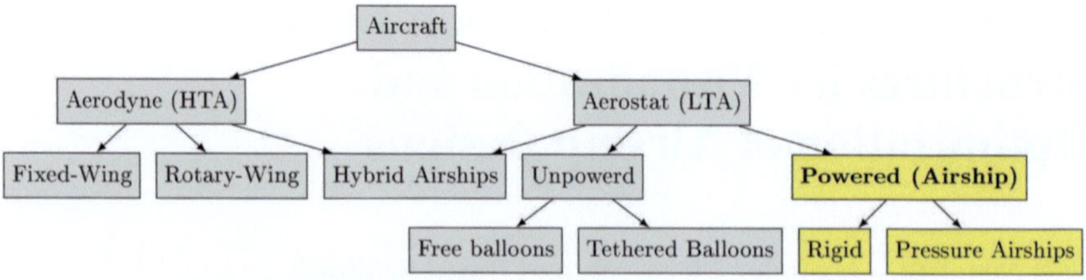

Fig. 1 Overview on categories of aircraft [1]

1.2 Objectives

The first objective is to store airship data in a data structure being able to visualize the dirigible easily and visually appealing. Also, the geometry should be parametrized in order to perform quick changes in the geometry and to calculate properties of preliminary designs. Consequently, this data is used for optimization in early design phases. Optimizing one parameter with a simple cost function is the second objective of this work.

2 Problem Definition and Formulation

First steps in designing an airship are to define its principal characteristics to fulfill certain requirements [2]. Aircraft lighter than 15 t with less than 20 passengers may be certified as commuter aircraft [3], which comes with cheaper development and certification costs and should be this design problems driver.

The $(\frac{L}{D})$ of the airships hull (also often referred to as 'fineness ratio') influences both the weight and the aerodynamics of the aircraft significantly. Higher slenderness comes with the cost of added structural mass but influences the aerodynamics positively.

The considered design optimization problem can thus be summarized as 'Finding the optimal $\frac{L}{D}$ for an airship fulfilling the requirements of the commuter category'.

3 Methodology

This section provides details about airship modeling and the performed calculations for estimation of parameters. The model and estimated values are then used in the design optimization.

3.1 Airship Modeling

Optimizing the geometry of an aircraft requires a parametric model in order to perform automated changes of the geometry. Solving the problem defined in Sect. 2 can be done using CPACS files and methods for the automated generation of varying hull and stabilizer geometries.

Central model approach Storing airship data in a centralized model is possible using CPACS [4]. CPACS is a XML-based structure developed by the DLR. It stores the parametric description of airplane and rotorcraft geometries and several other parameters such as mission definitions or the inputs and results of various analyses. The idea is having one centralized model as shown in Fig. 2 that is used for different applications. Centralized models are already established in different fields and have been used by airplane and helicopter manufactures for decades.

Using the CPACS data schema has the benefit of the existence of extensive libraries that come with the data structure. When using a CPACS file following the standard XML schema by the DLR, the TiGL Geometry Library (TiGL) C++ libraries and associated Python, MATLAB and Java bindings use the parametric description in the file for full three-dimensional visualization. The TiGL Libraries offer also other functionalities for modification of CPACS files and computation of geometric properties like surface area, volume or largest diameter [5].

For demonstration of the worthiness of a parametric description of airships using a CPACS-file, a historic airship with an actual flight record has been reengineered; see Fig. 3.

Hull geometry modeling In search for the perfect submarine shape, L. Landweber and M. Gertler developed a mathematical description of aerodynamically optimized

Fig. 2 Common Parametric Aircraft Configuration Schema (CPACS) method [4]

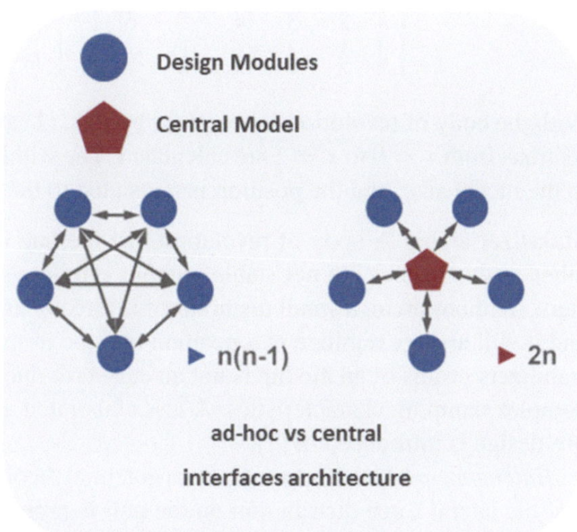

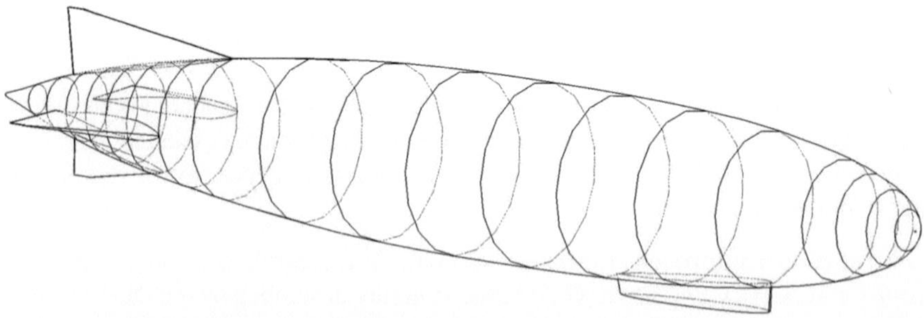

Fig. 3 CPACS geometry from a reengineered LZ 120. The figure shows a geometry that is exported to the IGES format and opened with CAD software

bodies [6]. Describing the shape of the bodies needs five parameters such as $\frac{L}{D}$, prismatic coefficient c_p, location of maximum thickness m, bow- and stern-radii r_0 and r_1 [1]. The equation describing the bodies shape as a function of longitudinal distance x is a polynomial of the sixth order.

$$f(x) = \left(\frac{L}{D}\right)^{-1} \cdot \sqrt{a_1 \cdot x + a_2 \cdot x^2 + a_3 \cdot x^3 + a_4 \cdot x^4 + a_5 \cdot x^5 + a_6 \cdot x^6}, \quad (1)$$

where a_1 to a_6 can be solved with the formulations of the four other shape parameters that are not used in Eq. (1). Summarized, this equals to

$$\begin{bmatrix} 1 & 0 & 0 & 0 & 0 & 0 \\ 1 & 1 & 1 & 1 & 1 & 1 \\ m & m^2 & m^3 & m^4 & m^5 & m^6 \\ 1 & 2m & 3m^2 & 4m^3 & 5m^4 & 6m^5 \\ 1 & 2 & 3 & 4 & 5 & 6 \\ \frac{1}{2} & \frac{1}{3} & \frac{1}{4} & \frac{1}{5} & \frac{1}{6} & \frac{1}{7} \end{bmatrix} \cdot \begin{bmatrix} a_1 \\ a_2 \\ a_3 \\ a_4 \\ a_5 \\ a_6 \end{bmatrix} = \begin{bmatrix} 2r_0 \\ 0 \\ \frac{1}{4} \\ 0 \\ -2r_1 \\ \frac{1}{4}c_p \end{bmatrix}. \quad (2)$$

With the body of revolution shape given by Eqs. (1) and (2), radii from a number of sections from $x = 0$ to $x = 1$ are calculated. The width of each section is then scaled to the model size, and the position is translated to the overall length of the model.

Stabilizer sizing A body of revolution like the one formulated in Sect. 3.1 and all other airship bodies are not stable without stabilizers as counteracting areas at the stern. Without them, a small disturbance is already enough to turn the airship body, and it will always stabilize in a position oblique to the direction of flow. Sizing the stabilizers or fins of an airship is not an easy task due to the inherent instability and complex damping characteristics. A less elaborated approach which is sufficient in pre-design is introduced in [7].

Hull moment Munk's approach from potential theoretical flow simulation assumes that the lateral force distribution on the hull is proportional to the derivative of the cross-sectional area in cross flows [7].

The moment M_{hull} at the position of maximum thickness of the body is described with the density of the surrounding fluid ρ, the velocity v, the overall volume V, the angle of attack α and two factors k_1 and k_2, describing the additional forces from the movement of the hull. The moment equals to

$$M_{\text{hull}} = \frac{\rho}{2} \cdot v^2 \cdot V \cdot (k_2 - k_1) \sin 2\alpha. \tag{3}$$

The angle of attack is chosen to be 5% because drivers of the stabilizers sizing are small obstructions while cruising. The k_1 and k_2 factors are being summarized to the Munk factor k that can be read from a table depending on the diameter over thickness ratio ($\frac{D}{L}$), reciprocal of $\frac{L}{D}$, found in [8].

Stabilizing moment The induced moment from the hull requires a contradicting moment: the moment M_{stab} induced by the stabilizer. The equilibrium of moments is calculated with the distance from the center of volume to the so-called 25% chord line of the stabilizer l_{stab} and the induced lateral force by the stabilizer F_{stab}:

$$\begin{aligned} M_{\text{hull}} &= M_{\text{stab}} \\ &= F_{\text{stab}} \cdot l_{\text{stab}}. \end{aligned} \tag{4}$$

A relative value for the lever arm l_{stab} depending on the overall length of the airship and a fixed elongation of the stabilizers Λ enables the calculation of lift curve slope $\frac{\delta C_a}{\delta \alpha}$ after the theory of wings with small elongation [9]. Further, the force F_{stab} required for equaling stabilizer moment and hull moment is calculated from the projected area $A_{\text{stab,projected}}$ of the stabilizer:

$$\frac{\delta C_a}{\delta \alpha} = \frac{2 \cdot \pi \cdot \Lambda}{3 + \Lambda} \tag{5}$$

$$F_{\text{stab}} = \frac{\rho}{2} \cdot v_{\text{cr}}^2 \cdot \frac{\delta C_a}{\delta \alpha} \cdot \alpha \cdot A_{\text{stab,projected}}. \tag{6}$$

3.2 Parameter Estimation

In this section, we show drag and mass calculations using simplified assumptions. All estimated parameters are being saved in the central CPACS model.

Drag Estimation The drag force of the airship in stationary flight with the cruise speed of 80% of the maximum speed is used as reference drag. The flow is assumed to be fully turbulent, which is a conservative assumption. Laminar flow is, in all cases unlikely, looking at high Reynolds numbers being found in airship aerodynamics.

The total drag is reduced to two parts: the drag induced by the hull and the stabilizers.

Hull The calculation of the drag is done by a number of steps, from which the first is to determine the actual Reynolds number Re. Cruise speed, v_{cr}, over all airship length L_{OA} and the kinetic viscosity of the surrounding air in International Standard Atmosphere (ISA) ν are used to calculate:

$$\mathrm{Re} = \frac{v_{cr} \cdot L_{OA}}{\nu}. \tag{7}$$

The empirical approach by Prandtl's boundary layer theory gives a frictional drag coefficient value of:

$$c_{w,friction} = \frac{0.455}{\log(\mathrm{Re})^{2.58}}. \tag{8}$$

A frictional drag force $F_{w,friction}$ is given by the hulls wetted surface area A_{hull} as reference area, cruise speed v_{cr}, and ambient air density ρ. With the before calculated $c_{w,friction}$, we get

$$F_{w,friction} = \frac{\rho}{2} \cdot v_{cr}^2 \cdot c_{w,friction} \cdot A_{hull}. \tag{9}$$

Applying a form factor based on geometry of rotational bodies, we can now calculate the total drag after Hoerner [10] with the reciprocal fineness ratio $\frac{D}{L}$ as follows:

$$F_{w,hull} = F_{w,friction} \cdot \left[1 + 1.5 \cdot \left(\frac{D}{L}\right)^{\frac{3}{2}} + 7 \cdot \left(\frac{D}{L}\right)^3 \right]. \tag{10}$$

Stabilizer Stabilizers contribute the largest share of drag after the hull itself. The reference area is the wetted surface area of the stabilizers A_{stab}, and the drag coefficient $c_{w,stab}$ is estimated to be 0.1. The drag force $F_{w,stab}$ induced by the stabilizers is then

$$F_{w,stab} = c_{w,stab} \cdot \frac{\rho}{2} \cdot v_{cr}^2 \cdot A_{stab}. \tag{11}$$

Total drag The total drag $F_{w,tot}$ is the sum of the single shares. Gondola and other extensions are being neglected, and only an interference share of 3% is added to the drag. The total drag is given by

$$F_{w,tot} = 1.03 \cdot (F_{w,hull} + F_{w,stab}). \tag{12}$$

Mass Estimation Aircraft masses are classified into several categories. This section provides the classification used for the design optimization.

Operating Empty Mass The operational empty mass (m_{OEM}) includes all masses of the airship except fuel mass and payload which are recorded separately. We use Normand's scaling method for estimation of the m_{OEM}.

Burgess describes the application of Normand's equation to estimating the sizes and weights of airships [2]. Here, the mass of an airship design is divided into 14 weight groups. With the fixed characteristics and independent variables that need to be assumed by the designer, the mass of each weight group can be scaled by the dependent variables.

J. Eissing further improved the approach by including more dependent variables [11]. He also adapted Burgess approach to a selection of real-world airships [12]. Following his approach and taking the scaling parameters calculated by averaging the real-world airships, weights from the 14 weight groups can be estimated for a given hull shape with few parameters.

The individual masses of the weight groups m_i sum up to the total mass

$$m_{\text{OEM}} = \sum_{i=1}^{14} m_i. \tag{13}$$

Fuel mass The assumptions made for calculating the fuel mass are that the airship travels with a constant cruise speed ISA at sea level. Contradicting airplanes fuel consumption, a near equilibrium airship does not need a lot of additional power for starting, making the assumption more valid. Also, the simulation is simplified to a single powertrain including motor, drivetrain, gears and rotor.

Aerodynamic power P_{aero} is given by

$$P_{\text{aero}} = F_{\text{w,tot}} \cdot v_{\text{cr}}. \tag{14}$$

and the ratio of aerodynamic power and shaft power (delivered power) according to [13] by

$$\eta_{\text{delivered}} = \frac{P_{\text{aero}}}{P_{\text{shaft}}}, \tag{15}$$

Calculating P_{shaft} with Eq. (15) and adding an overall drivetrain efficiency $\eta_{\text{drivetrain}}$ gives the required motor power that is multiplied with the trip duration t_{cr} and a specific fuel consumption of the motor $c_{\text{fuel,cr}}$ to get the fuel mass m_{fuel} with

$$m_{\text{fuel}} = P_{\text{shaft}} \cdot \eta_{\text{drivetrain}} \cdot c_{\text{fuel,cr}} \cdot t_{\text{cr}}. \tag{16}$$

This mass is assumed as a static mass, whereas a more detailed simulation would respect the mass loss due to fuel burn.

Payload Payload can be cargo of different kind or passengers. The monetary value of the payload highly depends on the kind of cargo that is being transported. The payload m_{payload} is the mass that remains when subtracting all other masses from the total mass m_{TOT} of the airship:

$$m_{\text{payload}} = m_{\text{TOT}} - m_{\text{OEM}} - m_{\text{fuel}}. \tag{17}$$

3.3 Design Optimization

Optimization problems are best approached with a structured method. MDO is a systems engineering approach where a number of disciplines are considered for solving design problems. Using CPACS has the advantage that the central-based approach of the data format can be used in different tools of the MDO. Solving airship design problems using MDO enables designers for a fast evaluation of different designs.

Set up Simplicity and reproducibility are the driving forces in this optimization setup. Using a single free variable as input of the design helps in visualizing the problem and making it comprehensible. CPACS is a geometrically driven data format and offers parametric description of geometries, thus choosing a geometric parameter for a variation of the designs is consistent. Using an algorithm for creating Gertler shapes, following Sect. 3.1 enables simple variation of the parameters used.

Variable parameter The fineness ratio is an important characteristic influencing the appearance, weight and drag of an airship significantly.

Using the algorithm for the creation of hull shapes, we created a set of shapes with varying $\frac{L}{D}$ from 5 to 10 and a step width of 0.1 (Fig. 4).

Fixed parameters There are a number of fixed parameters. Table 1 lists the most important fixed parameters. The total mass is chosen at $0.98 \cdot 15\,t$, which is an important limit for the certification of aircraft. Airships below 15t carrying less than 20 passengers may be certified as 'commuter airships'.

The cruise speed is a fraction from the maximum speed. 70knots is the airspeed just above 12beaufort which the structure of the airship must withstand while moored to the mast. Thus, designing an airship with lower maximum speed does not reduce structure mass.

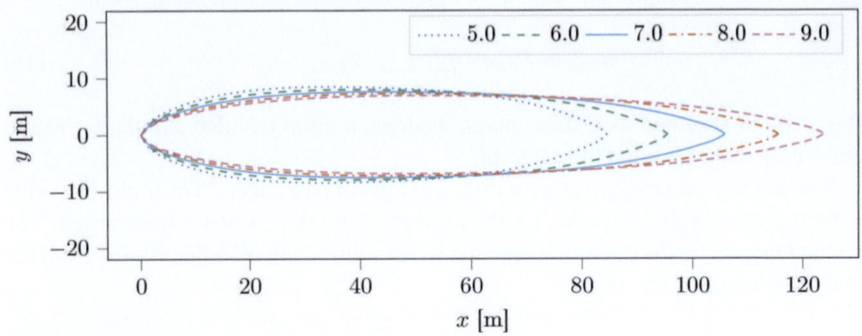

Fig. 4 Variation of the hull shape with varying fineness ratio

Table 1 Selection of important fixed parameters used in the design optimization

Parameter	Symbol	Value	Unit
Total Mass	m_{tot}	14.7	t
Number of gas cells	n_{GC}	12	–
Number of cross sections	n_{sec}	50	–
Cruise speed	v_{cr}	28.8	$\mathrm{m\,s^{-1}}$
Cruise time	t_{cr}	8	h
Specific fuel consumption	$c_{fuel,\,cr}$	1.4	$\mathrm{kg\,h^{-1}\,kW^{-1}}$

Cost function The cost function must depend upon the variable parameters and have a non-discrete number as function value. The cost function calculates the payload as the results of the airships total mass less m_{OEM} and m_{fuel}, which depends on the drag. Both values drag and m_{OEM} depend on $\frac{L}{D}$.

$$m_{payload} = f\left(\frac{L}{D}\right)$$
$$m_{payload} = m_{tot} - m_{OEM}(\frac{L}{D}) - m_{fuel}\left(\frac{L}{D}\right). \tag{18}$$

Optimizer The type of optimizer chosen for the given optimization problem is a brute force (BF) optimizer. BF optimization is a method where the cost function is evaluated at each of a given number of points. The optimal design is then found by choosing the *maximum* value from all evaluated cost function points. This method is connected to a high demand of hardware resources and needs more calculation time for complex problems than other methods, but is a fast and simple approach for solving optimization problems of lower complexity. The optimizing algorithm can be simplified by the steps in Algorithm 1.

Algorithm 1 Brute force optimization.

for $\frac{L}{D} = 5, \ldots, 10$ **do**
  copy CPACS xml file
  create gertler shape, size fins
  calculate drag, m_{fuel}, m_{oem}
  $\mathbf{m}_{payload} \leftarrow m_{tot} - m_{OEM} - m_{fuel}$
end for
OPTIMUM $\leftarrow$ MAX($\mathbf{m}_{payload}$)

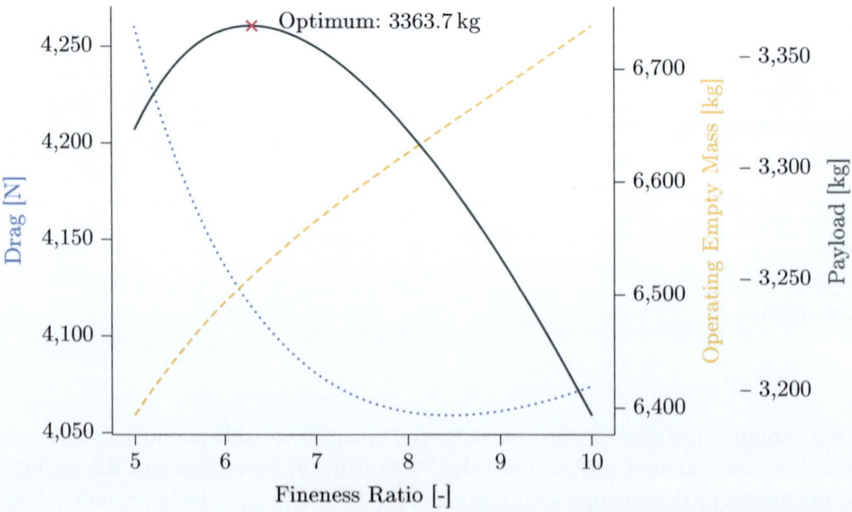

Fig. 5 Resulting payload and the counteractive values for operating empty mass and drag

4 Results

The airships drag force is higher at lower $\frac{L}{D}$ values, which represents a basic rule in aerodynamics: Slender bodies show lower drag. The drag is rising again at above 8.5 because of frictional resistance. Slender bodies do have a larger wetted surface when we keep volume constant and more wetted surface equals more frictional resistance.

The counteracting effect is the airships mass. Here, we see an almost linear rise of structural mass over fineness ratio. The sphere is the optimal shape in terms of surface or structure needed to encase a volume. Deviation from the structural 'perfect' shape, in this case represented by slenderness, results in higher structural weight.

The results in Fig. 5 emphasize the counteracting effects driving the design optimization and the resulting payload. The optimum at $\frac{L}{D} = 6.3$ gives a payload of 3363.7 kg.

5 Discussion and Conclusions

With the methods used, this work introduced a common data structure for airships and further demonstrated the benefits of the data structure by performing a MDO. Further, the designs can be visualized from the parametric descriptions. The results are reproducible and met expected performance despite major simplifications in the methods used.

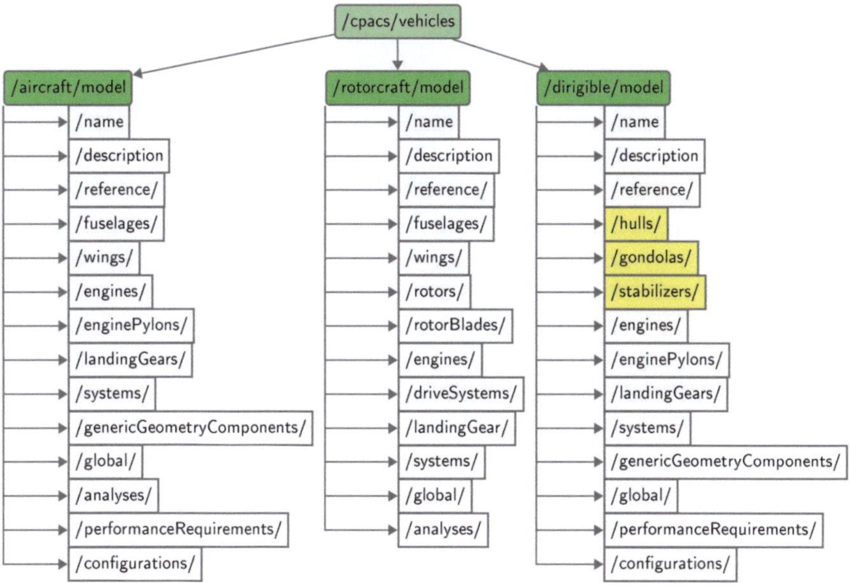

Fig. 6 A cutout from a CPACS schema showing the attributes and elements of the three vehicle elements. Elements, that are not a copy of aircraft or rotorcraft elements, are marked in yellow. Attributes can be recognized by the absence of a slash behind their name, whereas elements names are followed by one, indicating that they have subordinate attributes and/or elements

We have two proposals for future work: first, an expansion of the underlying CPACS open-source schema shown in Fig. 6 and second several ideas for the further development of the design problem optimization:

- Adding more optimization variables.
- Using more elaborated solvers, preferably gradient based.
- Setting up a constraint optimization where the airship has to fit into a hangar.

References

1. Eissing J, Design considerations for cargo airships (15 Mar 2019)
2. Burgess CP (2004) Airship design. University Press of the Pacific, Honolulu
3. Federal Aviation Administration: LFLS: Airworthiness requirements for the type certificate of airships in the categories normal and commuter. https://www.faa.gov/aircraft/air_cert/design_approvals/airships/airships_regs/media/aceAirshipLFLS.pdf (13 Apr 2001). Accessed 09 Dec 2021
4. Alder M, Moerland E, Jepsen J, Nagel B (2020) Recent advances in establishing a common language for aircraft design with CPACS. In: Aerospace Europe conference 2020

5. Siggel M, Kleinert J, Stollenwerk T, Maierl R (2019) TiGL: an open source computational geometry library for parametric aircraft design. Math Comput Sci 13(3):367–389. https://doi.org/10.1007/s11786-019-00401-y
6. Landweber L, Gertler M (1950) Mathematical formulation of bodies of revolution. Washington
7. Munk M (1979) The aerodynamic forces on airship hulls
8. Funk TL, Wagner S (2002) Experimentelle untersuchungen von rumpf-leitwerk interferenzen bei luftschiffen. In: Deutscher Luft- und Raumfahrtkongress (ed) Tagungsband Deutscher Luft- und Raumfahrtkongress 2002 (23–26 Sept 2002)
9. Frank P (1990) Die Auslegung von Flugzeugen mit geringstem Antriebsleistungsbedarf
10. Hoerner SF (1965) Fluid-dynamic drag: practical information on aerodynamic drag and hydrodynamic resistance. Hoerner Fluid Dynamics, 16 Dec 2021
11. Eissing J (2015) Normand scaling tool
12. Lancaster JW, Feasibility study of modern airships, phase 1. vol. 4: Appendices. https://ntrs.nasa.gov/api/citations/19750023962/
13. Eissing J (2003) Widerstand und propulsion von luftschiffen

Airship Sling-Load Operations: A Model Flight-Test Approach

Johannes Eissing, Carl Nenoff, Erich Fink, Martin Zobel, and Florian Antrack

1 Introduction

This section first provides background on the topics being discussed in this paper and continues with the objectives.

1.1 Background

Backgrounds crucial for following the conducted research is given in this section. It starts with technical background on airships, followed by operational backgrounds on aircraft cargo, and concluding with background on the control and operation of airships.

Airships Airships, as a means of transportation, are not a new idea. Historic examples showed that high payloads can be reached with significantly higher transport efficiency compared to airplanes and rotorcraft [1]. Vertical Take-Of and Landing

J. Eissing (✉) · C. Nenoff · E. Fink · M. Zobel · F. Antrack
Aerarium Luftschifftechnik e.V., TU Berlin, Marchstrasse 12-14, 10587 Berlin, Germany
e-mail: johannes.eissing@aerarium.de
URL: http://www.aerarium.de

C. Nenoff
e-mail: carl@aerarium.de

E. Fink
e-mail: erich.fink@aerarium.de

F. Antrack
e-mail: florian.antrack@dlr.de
URL: https://www.dlr.de/EN

F. Antrack
DLR, German Space Agency, 53227 Bonn, Germany

D. Shukla (ed.), *Lighter Than Air Systems*, Lecture Notes in Mechanical Engineering,
https://doi.org/10.1007/978-981-19-6049-9_3

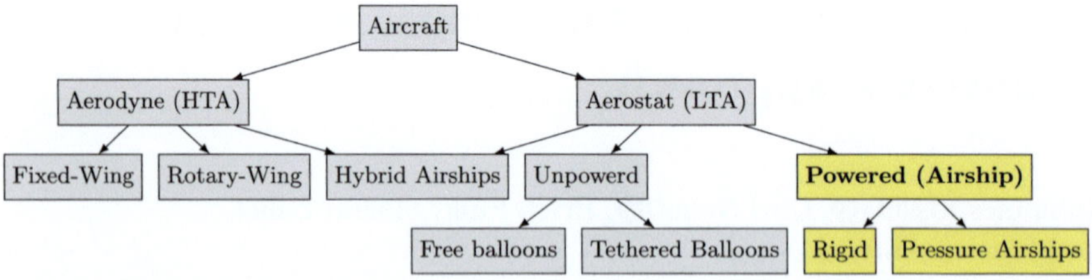

Fig. 1 Overview on categories of aircraft [2]

(VTOL) capabilities are also given with Lighter-Than-Air (LTA) systems. Reducing emissions is a focus of this century, and something airships can help to achieve. Figure 1 shows the taxonomy of aircraft.

Near equilibrium airships There are several classes of LTA vehicles, portrayed in Fig. 1. This research does only consider dirigibles, which means powered, near equilibrium aircraft, called airships. The considered subclass of aircraft is indicated in Fig. 1 by yellow color. Near equilibrium is the capability of achieving zero static heaviness during normal flight operation, meaning that such airships float in the air like fish or submarines in water. An airship heavier than the displaced air sinks to the ground, an airship lighter than air rises until it reaches the height of aerostatic equilibrium.

The inherent safety feature of near equilibrium airships facilitates certification. Even an 'all engines inoperative' incident leaves the airship safe and controllable as a balloon. The consequences of an airship flying in balloon mode are far from the consequences of heavier than air vehicles urgent need for a landing site. Because of this inherent safety feature and the low noise emission, airships could be the preferred air transport system for flights over sensitive areas like areas with a high population density or integrated natural reserves.

Static heaviness A certain degree of deviation from the equilibrium weight is tolerable and can be trimmed using ballast and aerodynamic effects. This tolerable degree is referred to as the static heaviness. Static heaviness for near equilibrium airships has been demonstrated ranging from -5 to $+10\%$ of the airships design equilibrium weight. Static heaviness will be used in this document as a reference to categorize the different types of cargo.

Cargo Cargo is transported goods onboard of vehicles. Different kind of cargo needs to be described in order to understand the transport problem [2]. Due to the nature of LTA technology, the types of cargo can be referred to the static heaviness of an airship. Bottlenecks for transportation are load exchange and maneuverability during the load exchange procedure.

General cargo Pallets, barrels, or bags in larger dimensions, and also standardized container frequently used in aviation, are typical break bulk cargo. Single bits of cargo are usually within the static heaviness of an airship. Single cargo items lighter than the static heaviness can be loaded or re-loaded unit by unit. Each unit dropped needs to be replaced by a corresponding cargo or ballast mass. Loading and unloading of such cargo is common practice in LTA operations.

Passenger One of the 'cargo' types that is currently transported commercially successful via airships are passengers. *Zeppelin NT* flies several airships in collaboration with the Commercial Air Transport (CAT) operators *Deutsche Zeppelin- Reederei (DZR)* and *Michelin* in Friedrichshafen.

Passengers are from the loading procedures point of view similar to general cargo. They are usually lighter than the static heaviness of an airship. Passengers can easily be exchanged with other passengers or ballast. Standard procedures exist. Their *'self-loading capability'* is an advantage against general cargo. There are on the other hand disadvantages like comfort and safety requirements, that have a strong impact on aircraft certification and thus on direct operating costs.

Bulk cargo Cargo that does not come in discrete numbers or packet numbers is referred to as bulk cargo. A continuous process of loading or unloading and ballasting is necessary. Three different types of bulk cargo can be identified.

First, there is gaseous cargo, for which there are no procedures existing today that consider the special requirements coming with handling mostly flammable gaseous cargo. Still, transporting natural gas or hydrogen via airship is an interesting topic and might become important.

The second kind of bulk cargo is so-called dry trade like gravel, ore or agricultural bulk solids like grain.

'Wet trades' like oil, agricultural fluids or fuel are probably the most established bulk goods, as refueling is a necessary part of operation. Procedures have been developed both for moored and untethered 'in-flight' refueling.

Heavy loads Loads that excessively exceeds static heaviness are defined as heavy loads. In recent airship projects, special interest has often been transportation of heavy loads. Heavy loads are often outsized and strain existing infrastructure due to high punctual forces and larger than usual dimensions. LTA operations could and should be a viable solution for heavy loads, especially in remote areas without or with insufficient infrastructure. Examples of heavy loads are large wind energy converter compounds or surgical wood extraction from forests.

Operation modes Both the aircraft control and the control of load hooks have different states, referred to as 'flight modes' and 'hook modes'.

Flight modes Modern airships often use different flight control modes, 'cruise mode' during free flight and 'hover mode' during landing and loading operation.

During cruise mode, efficient power usage and stable flight are the most important goals. They are being achieved by properly designed stabilizers and powertrain of the airship.

Hover mode requires the airship to hold a precisely controlled position. Landing or masting and un-masting operations, but also sensor measurement missions and loading operations are driven by position control, which is crucial during those operations.

There are different concepts of how the two flight modes are being implemented, but all modern airships distinguish the two modes. Depending on the deployment scenario of an airship design, emphasis on one of the modes determines an airships propulsion configuration.

Hook modes Helicopter cargo hooks function as 'cargo release units', allowing for remote controlled cargo release. Such systems, mostly of electromagnetic type, are certified for use in Helicopter External Sling-Load Operation (HESLO) [3]. Airship External Sling-Load Operation (AESLO) is going further than HESLO with two exchanged loads instead of one and in consequence two hooks that need to be controlled separately. Introducing two hook modes is important to address certification requirements.

1. Hook mode 1: Load 1 stays
2. Hook mode 2: Load 2 stays.

At some point during loading when both loads are loaded, the Load Master onboard the airship or on ground decides to switch the hook mode. Doing this, free flight condition can always be achieved in a short amount of time.

1.2 Objectives

Objective of this research is focused on how load exchange with near equilibrium airships can be performed. To the knowledge of the authors, proposed solutions aiming at similar targets have not been demonstrated successfully yet, or did not meet certification requirements. We see different viable approaches and want to demonstrate one promising idea on how to do load exchange operations beyond allowed static heaviness viable for airships.

The research should also identify weaknesses and critical points in load exchange procedures.

2 Problem Definition and Formulation

Especially loads excessively beyond static heaviness are of interest to us. Heavy loads introduced in part 1.1 do always require a special load exchange and may at the same time influence the airships' maneuverability negatively. Load exchange procedures have been proposed before, for example in two patents of the company CargoLifter [4, 5], but they potentially fail to correspond certification.

Furthermore, heavy loads often exceed an aircraft storing room dimensions. The solution of external transport may affect maneuverability negatively.

2.1 Load Exchange and Certification

The idea of loading LTA is to keep the Operating Mass (OM) at the same level at all times during operation. In particular, OM should not change beyond allowed static heaviness. When loads exceeding this limit are being dropped or taken, a load with the same mass needs to be exchanged with that load. With heavy loads, the question arises of how this should be achieved. Dropping a very heavy load consequences in a skyrocketing airship which becomes a large threat for pilots, crew and air traffic. Mitigating said risk is aim of this research.

Airship certification does also require to avoid the opposite state, that an airship is bound to the ground by large masses. The original citation of the certification specification is: *'During any cargo exchange [...] the airship must be capable of achieving a safe free flight condition within a time period short enough to recover from a potentially hazardous condition'* [6].

There are two cases, where aborting the loading procedure and achieving free flight condition as a safety measure against hazardous condition seems difficult. The first case is, when the ballast as well as the cargo are loaded onto the airship at the same time. The airship is only able to achieve free flight condition after loosing one of the two, which seems difficult to achieve in a short period of time.

The second case is when neither ballast nor cargo is loaded, and the airship is leashed to the ground like proposed in CargoLifters' patent [5]. Cutting the ropes in a hazardous situation might get the airship safe from the hazard, but achieving free flight while being significantly lighter than air is not possible and poses a great risk and danger as well.

2.2 Maneuverability

Aim of this research is to improve the maneuverability during load exchange operations. Maneuverability reduces risks in ground operations as well as the size and thus cost of the ground crew. Pilot workload is lessened, which is required by certification specifications.

Classical airships like the Zeppelin LZ-120 'Bodensee' required large numbers of ground crew [7], for it is low speed maneuverability was not sufficient to direct the airship towards the wind. Contemporary blimps like the ABC A60, A150, and A170 are smaller, but still need dozens of ground grew. More modern blimps like the Skyship 500 and Skyship 600 feature vectored thrust. This makes them VTOL capable, but still ground crew is needed to direct the airship toward the wind, as low speed rudder efficiency is not sufficient. Due to a stern lateral propeller and vertical propeller, the Zeppelin NT can direct itself toward the wind in low speeds. Dynamic positioning with no speed at all still remains a challenge.

Fig. 2 Onboard Systems' MD500 Dual Cargo Hook kit with Y-rope [8]

3 Methodology

Two steps will demonstrate development of systems and operations of a load exchange procedure for heavy loads with near equilibrium airships. First, a sling load operation is introduced. We then proceed by performing and testing the operational handling. For this purpose a specially developed and built airship model will be used.

3.1 Sling-Load Operation

The well known and often performed procedure Helicopter External Sling-Load Operation (HESLO) is being explained, and its working principal will be transferred to Airship External Sling-Load Operation (AESLO).

HESLO Loads, that cannot be carried inside a helicopter because of their dimension, position on ground or weight, are being carried externally using hooks and sling leashes. Loading and unloading of the load does not require the helicopter to land and thus happens in hover. Outside the rotorcraft, structural elements connect one or two cargo hooks or hoists operated by the pilot [3]. Figure 2 shows an example of such helicopter hooks.

Fig. 3 A Sikorsky Skycrane carrying an accommodation

The lift of a rotorcraft is exclusively aerodynamic lift from the rotor. As long as a load does not exceed the maximum payload of a given vehicle it is capable of carrying it. HESLO is a well established procedure with many applications. Figure 3 shows an eccentric example of using HESLO.

AESLO AESLO is the proposed and favored term of an operation that is oriented on HESLO, but considers the characteristics of airships. Two sling-loads and cargo hooks are what is needed when performing an AESLO. Both, delivery load and return load, are being carried by slings. In advance of an AESLO two loads have to be prepared, one attached to the airship and the other waiting on ground.

The proposed cargo loading and unloading operation consists of the following steps and is visualized in Fig. 4:

1. Airship approaches in cruise mode with two slings, hook 1 holds the delivery cargo. Hook mode 1, described in Sect. 1.1, is active.
2. Airship arrives at the pick-up destination and changes from cruise mode to hover mode.
3. In hover mode, dynamic positioning is used to place the airship precisely over the spot where the delivery load is to be placed. The load is placed on the ground, and the airship moves over the return load.
4. Return cargo is being loaded by the empty sling, held by hook 2.
5. Change hook mode. Hook mode 2 is active.

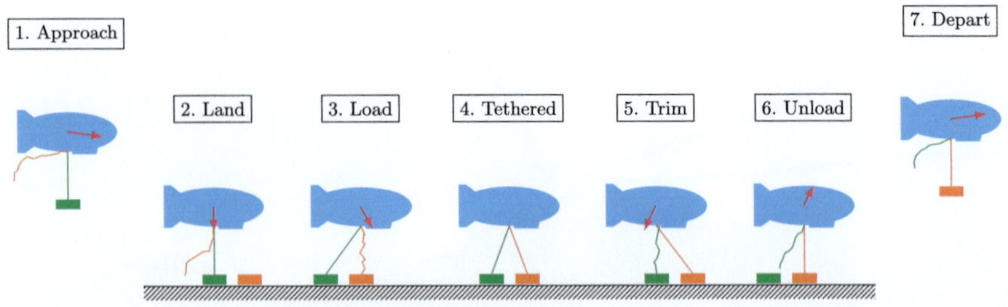

Fig. 4 Step-by-step AESLO operation

6. The delivery cargo is being unloaded and disconnected from the sling.
7. Airship changes from hover mode into cruise mode again, and leaves the loading site.

A critical point in load exchange operation is the certification requirement described in Sect. 2.1. To ensure the safety of the Operation and optimize the time until cruise mode is regained, a Load Master must monitor the load exchange and switch the hook modes.

In case of an event that makes evacuation necessary while both loads are loaded, one of the loads needs to be disconnected from the airship. During load exchange, the 'Load Master', introduced in Sect. 1.1, decides which load to cut in case of emergency. It is also the Load Masters responsibility that one load is connected to the airship at all times.

It is crucial for this operation that either the delivery or the return load is attached at all times. During load exchange, it is the Load Masters responsibility to decide, which load is to cut in case of a hazardous event by controlling the hook modes.

3.2 Model Flight-Test

Model flight tests allow statements about the performance of an aircraft in early design phases. In this section, similarity principles in smaller scale model testing are mentioned. It is continued with the test set-up and preliminary description of the test model to be built. The section concludes by describing the test runs, using the test set-up and airship model.

Similarity Principles Model test, in towing tanks, cavitation tunnels or in this case open environment, require special focus on the similarity principles. Froude's Law [9] gives a good estimation on the different units scaling. When building a model, geometric dimensions scale linear. Other dimensions however, like the time, angular rates, and frictional drag do not scale linearly. The units individual transformation from full size to model size needs to be considered in all model studies.

Fig. 5 Airship 'Starocore' at the fair and exhibition *Faszination Modellbau 2021* in Friedrichshafen. First AESLO Trials have been performed with this Airship model

Test set-up For the test procedure, a simpler set-up than described in Sect. 3.1 ensures conducting of the model test with focus on general feasibility. Concepts like the hook mode are named to prove the certification capability of the concept, but are not being investigated in this model flight test. Setting up a test for AESLO requires a model airship capable of precise positioning, carrying two attached cords below the center of buoyancy of the model. Furthermore, two loads of equal mass are required. The loads are attached to cords via manually operated hooks.

Model Main object of the AESLO model flight test is an airship model. The model should have enough buoyancy to carry a payload that exceeds the static heaviness significantly. Further, the model needs to have precise positioning control when in hover mode.

Propulsor configuration Experiences with airship models have shown that a propulsion set-up with four thrusters on pivoted pods can fulfill this requirement [10]. Figure 5 shows the model airship *Starocore* that has been built by this principle. The four pivoting thrusters are clearly visible in the Figure. The model showed good maneuverability in both hover and cruise mode but does not have enough volume to carry additional loads that exceed static heaviness.

Control Performing load operations requires hover control different from cruise control. This mode aims at easy precision hovering instead of stable flight. The different control modes can be implemented in the remote control running openTX. A switch is used to change flight modes. Figure 6 shows the controller layout on an openTX remote control for cruise and hover mode. We want to emphasize that the throttle in hover mode is being set permanently. This is necessary because an AESLO requires permanent thrust during the operation.

Test runs The model flight test follows AESLO from Sect. 3.1, displayed in Fig. 4. The total thrust vector of the four distributed propulsors of the airship is indicated by a red arrow (↗) indicating the estimated direction and magnitude of thrust. The steps from the figure in detail are the following:

1. Payload A is attached to airship via sling A. Airship approaches payload B in cruise mode and changes to hover mode.
2. Payload A is lowered to the ground at loading site of payload B.

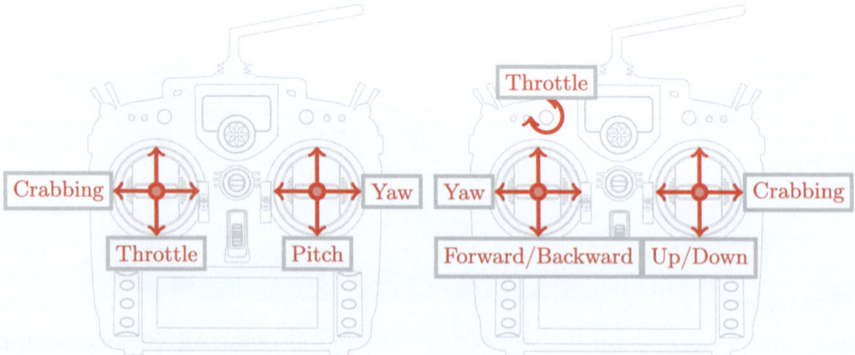

Fig. 6 Control sender configuration in cruise mode (left) and hover mode (right)

3. Airship moves over to payload B. Remote hook B is hooked to payload B.
4. Both slings are tense. Switching hook mode.
5. Airship moves toward payload A. Remote hook A is unhooked from payload A.
6. Airship moves over payload B.
7. Airship lifts off loading site. Flight mode change from hover mode to cruise mode.

Performing several recorded runs using an airship model generates the data necessary for evaluation of the proposed load exchange operation.

4 Results

A model as proposed in Sect. 3.2 has been built and named 'Hui Hui II'. Initial tests were performed in Berlin during the annual 'ELRT' event of the Technical University Berlin (TUB). The airship was built resembling its predecessor 'Hui Hui' which flew in Friedrichshafen in 2021 during the FAI CIAM Open International Competition F7B Model Airships. The models' configuration performed well under competition constraints, both in cruise and hover mode. This track record and the possibility of reusing electronic components lead to the resulting airship model. The Hui Hui II is, however, equipped with a larger envelope, while using the same propulsion system.

4.1 Cruise Mode

Handling qualities in cruise mode were as desired. Stability and controllability were perfectly given. The airship has been flown with a payload of 20% of the overall weight. When decelerating, the airship tended to pitch nose down. This tendency could easily be counteracted by elevator deflection.

Fig. 7 Close-up of the airship model used. The pig-leg configuration of the propulsion system allows dynamic positioning of the airship

4.2 Dynamic Positioning

Dynamic positioning capabilities are crucial for load exchange operations. At the same time, we had the largest difficulties with the airship operating in hover mode. This chapter discusses the controllability by degrees of freedom and proceeds with the operational observations made during the different phases of the flight test.

The two front propulsors provided constant thrust backwards and the rear propulsors provided constant forward thrust during hover mode. This 'pre-tension' provides quicker change in forward-backward direction, as compared to reversing the thrust of all four propulsors (Fig. 7).

Controllability Controllability in all six degrees of freedom has been observed as follows:

Surge (forward/backward) In cruise configuration, the design direction of the propellers has been set to forward thrust at forward speed. In hover configuration, the front propulsors were not swiveled backwards by 180 °C, but operated in reverse RPM. That means the efficiency of the propellers is largely reduced, as the propeller blades operate with reverse camber, and reverse profile sections. As a result, the airship developed more thrust in forward direction than in backward direction, when the propellers were set to equal RPM.

Sway (crabbing, or side wards motion) Crabbing ability has been proven poor, due to the large lateral added mass, and cross flow drag.

Heave (up/down) Heave control was satisfactory, while the asymmetric propeller settings provided additional workload to the pilot.

Roll Roll control in hover mode in this propulsion configuration is not possible, while considered not necessary.

Pitch Pitch control has not been investigated in depths during these tests. However, the airship reacted to pitch-up and pitch down commands.

Yaw (heading) Yaw control was satisfactory, while considerably less agile than with the predecessor airship 'Hui Hui'. This is due to the larger momentum of inertia, and higher yaw-damping due to the larger size.

Sway-yaw coupling When attempting to sway (crab), the airship developed an additional uncommanded yaw motion. This tendency is assumed to result from both, the cross flow drag of the empennage, and from the lateral added mass of the empennage. The two effects provide a forward-backward non-symmetry, leading to said sway-yaw coupling both, in steady- and in accelerated motion.

The human-machine interface, namely the mixer programming of the remote control transmitter, proved to deserve more intuitively operable settings. Several 'modes', as known from model aircraft remote control transmitter settings, are possible. The main challenge in this particular case might be the two very different modes of operations, namely, cruise and hover. Additionally, airships behave considerably different as compared to airplanes and helicopters or multicopters.

Operational observations During the test phases we made the following observations (Fig. 8).

Phase 1., Approach The workload to the pilot was low in cruise. When decelerating, a nose pitch down tendency has been observed. However, without high pilot workload, this tendency could be counteracted. Keeping the airship in position was

Fig. 8 The landing site is indicated by an 'A' for airship and marked with a two by two meter square and a smaller one by one meter square allowing for a rating of maneuverability. Both cargo loads, one attached to the airships sling and one at the loading site, are equal in weight

only possible by high pilot workload. Reason is considered the above-mentioned forward-backward non-symmetry in thrust, as well as poor crabbing ability, and the sway-yaw coupling.

Phase 2., Land When touching the ground too fast, the sling occasionally became slack. This led to bouncing of the payload from the ground. In full scale, this behavior might lead to high slack loads in the sling.

Phase 3., Load When trying to hover over return load 2, it proved difficult to keep load 1 stuck to the ground. Load 1 tended to get dragged along, as it proved tricky to keep the desired sling tension. However, the 'Ground Crew' attached sling hook 2 to return load 2.

Phase 4., Tethered This was the easiest part. In full scale, the pilots' task was to keep the airship aligned against the prevailing wind direction. As in the model test there was no wind indoors, the pilot workload was close to zero.

Phase 5., Trim Same as in Phase 3., Load, the difficulty here was in keeping the desired sling tension. If the tension in sling 2 got too high, Load 2 got dragged along instead of staying in position.

Phase 6., Unload As soon as hook 1 was released, Load 2 tended to bounce.

Phase 7., Depart Comparable to Phase 1., Approach, the pilots' workload was low, as the handling qualities of the airship were uncritical.

5 Discussion and Conclusions

A load exchange procedure for near equilibrium airships has been proposed by the authors. The loads considerably exceed operational static heaviness. The procedure resembles proven HESLO, adapted to near equilibrium airship airworthiness requirements.

Three bottlenecks have been identified for such operations, namely, low speed maneuverability, involved systems, and the operational procedure itself. To comply with low speed maneuverability requirements, a propulsion configuration has been proposed that has proven successful in numerous model flight tests. The proposed load exchange system is an external two-sling configuration, as known from HESLO. We attempted to demonstrate the operational procedure in a model flight test program.

5.1 Discussion

The model flight test program indicated the general feasibility of the proposed approach. However, some challenges have been identified that must be addressed in further research:

Forward-backward non-symmetry of propulsor thrust in hover configuration The non-symmetry in forward-backward thruster configurations proved to increase the pilot workload. This can easily be addressed by, e.g., swiveling the forward thrusters backward for cruise mode. Variable-pitch propellers might also provide a solution.

Sling tension monitoring To prevent bouncing and dragging of the load, the sling tension must be monitored and controlled more closely. Hoists, operated by the load master might be a solution, but add complexity and weight. A spring-balance, visible to pilot and ground crew, might provide a pragmatic solution.

Sway-yaw coupling Yawing due to sway proved to increase pilot workload. However, this issue can be addressed even without an autopilot, by mixing laws for control inputs.

Human machine interface Many different modes are possible for inceptor arrangements. It might be advisable, though, to align with helicopter control input arrangements, as known from manned operations.

5.2 Conclusion

The feasibility of heavy cargo load exchange by near equilibrium airships opens a wide range of air cargo operations, where low speed is permittable. Significant fuel savings can be achieved, meaning drastic cuts in emissions, while increasing range and endurance considerably. In the same time near equilibrium airships can be operated with no downwash, making ground operations much easier and safer, providing enhanced acceptance by authorities and local residents. While challenges became evident during the model test program, pragmatic solutions are in sight, encouraging further investigations in this encouraging approach to future sustainable air transport.

References

1. Eissing J, Sustainable air transport by airship (10 Dec 2020)
2. Eissing J, Design considerations for cargo airships (15 Mar 2019)
3. Civil Aviation Authority (January 2021) Helicopter external sling load operations. https://publicapps.caa.co.uk/docs/33/CAP%20426%20Helicopter%20external%20sling%20load%20operations%20Edition%205.pdf. Accessed 14 Dec 2021
4. Hegner B (2006) Verfahren und anordnung zum lastaustausch für nicht gelandete luftfahrzeuge. https://patentimages.storage.googleapis.com/36/56/84/4e0414923c78d1/DE10148589B4.pdf. Accessed 09 Dec 2021
5. Schäfer I, Method for the precise setting down or picking up of cargo from airships: European patent. https://patentimages.storage.googleapis.com/a7/6c/75/cc6ba6d327a39e/EP0846071B1.pdf. Accessed 10 Dec 2021
6. Luftfahrt Bundesamt, Germany (March 2000) Transport airship requirements: Tar. https://www.faa.gov/aircraft/air_cert/design_approvals/airships/airships_regs/media/aceAirshipTARIssue1.pdf. Accessed 10 Dec 2021
7. Kleinheins P, Meighörner W (2007) Die großen Zeppeline: Die Geschichte des Luftschiffbaus. Springer, Berlin

8. Rotor & Wing International (2019) Onboard systems' md500 dual cargo hook kit certified by faa - rotor & wing international. https://www.rotorandwing.com/2019/02/26/onboard-systems-md500-dual-cargo-hook-kit-certified-faa/. Accessed 05 Jan 2022
9. Sogreah (2006) Similarity principles
10. Eissing J, Kämpf B (2002) Die antriebskonfiguration des cargolifter cl160. In: DGLR LTA Workshop V

Optimization of TPU/TiO_2 Films Using UV Additives for Improved Weather Stability of LTA Hull Materials

Bharti Rana, Neeraj Mandlekar, Shuchita Tomar, Mangala Joshi, and S. Wazed Ali

1 Introduction

Lighter than Air (LTA) systems work at very high altitudes in the stratosphere, where they face harsh weather conditions, thereby imposing significant challenges for the material to be used in LTA systems. The material should withstand damage due to high aerodynamic stresses [1]. The confrontation to hull bending moments that cause hog and sag moments in surface airships is deeply rooted in the strength of the fabrics. Also, it decides the largest feasible size of LTAs. Fabrics made of high-performance fibers, viz. Kevlar, Vectran, Zylon, spectra, etc., have been used, but the major limitation lies in their inferior UV performance that ultimately affects the life of LTAs. Hence, the UV performance of hull material is a big concern.

To overcome such challenges, various researchers have developed coated and laminated materials for aerostats and airships [2–6]. As high-altitude LTAs must stay aloft during their deployment required for mission persistence as well as to save time for inflation and deflation. Helium detainment in high-pressure airships is a tedious task as helium being a tiny molecule can diffuse through hull material very quickly. UV rays can generally be subdivided into three regions, as shown in Fig. 1 [7].

Various organic and inorganic UV shielding agents can be used considering the deteriorating effect of UV rays on polymers in coatings. Moreover, some studies reported the use of metallic and reflective coatings. The present UV barriers used are inadequate at higher altitudes. Inorganic UV protecting agents mainly consist of semiconductors such as TiO_2, ZnO, SnO_2, CeO_2, etc., as shown in Fig. 2, and these can be multifunctional in nature. These types of agents are also well known for their non-toxicity and eco-friendly nature as compared to organic ones [8].

B. Rana (✉) · N. Mandlekar · S. Tomar · M. Joshi · S. W. Ali
Department of Textile and Fibre Engineering, Indian Institute of Technology, Delhi 110016, India
e-mail: Bharti.Rana@textile.iitd.ac.in

D. Shukla (ed.), *Lighter Than Air Systems*, Lecture Notes in Mechanical Engineering,
https://doi.org/10.1007/978-981-19-6049-9_4

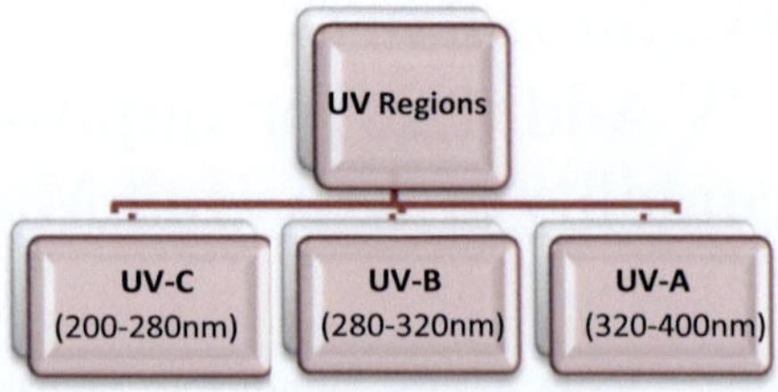

Fig. 1 Subregions of UV rays

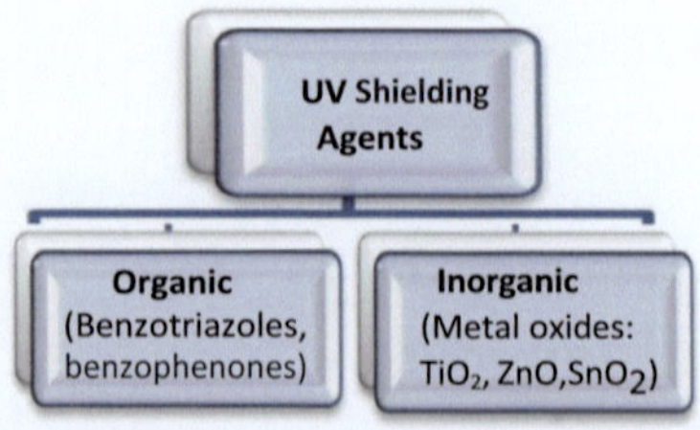

Fig. 2 Types of UV shielding agents

TiO_2 has gained much attention as it is non-toxic, chemical stable at high temperatures, and stable under UV exposure. It shows good scattering or reflecting properties due to its high refractive index. Additionally, it can also absorb UV rays as it is semiconducting in nature which is advantageous for aerostat applications [9]. Yang et al. have demonstrated the improved mechanism of TiO_2 as UV resistant additive [10].

Liu et al. have formulated the coatings consisting of Titanium dioxide and organic UV blockers for the application of UV protection in Vectran fibers to be used in airships. Classification of organic UV absorbers is shown in Fig. 3, and these generally support the polymer matrix to impart good UV resistance properties.

Benzotriazole (UV-1130) showed high UV absorbance in the range of 300–385 nm. Triazine (Tinuvin®477) UV absorber exhibits broad UV resistance in the UV range from 280 to 380 nm. Hindered amine light stabilizer (HALS) (Tinuvin®123) is supposed to be added along with other UV additives [11].

A similar work based on PU films consisting of antioxidants and composite stabilizers has been carried out by Wang et al. Antioxidant-based PU film was proven to provide anti-yellowing properties while composite UVAs-based PU film showed improved weathering properties toward UV and ozone [12]. Reflective metallic coatings have also been explored to be used as a UV-protective layer but limit their use

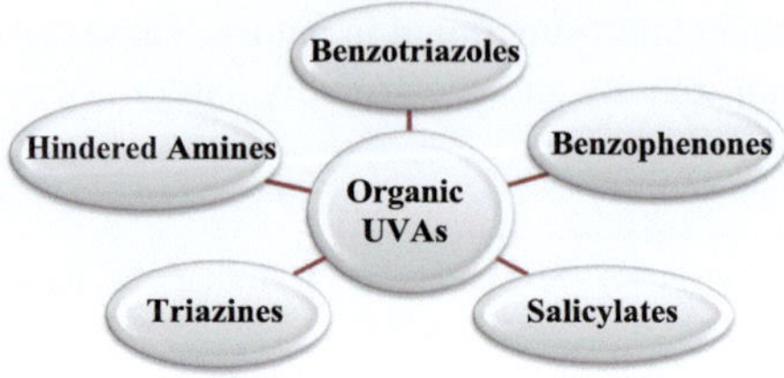

Fig. 3 Classification of organic UV absorbers used in coatings

due to adhesion problems and tear propagation over a long time. However, PU has been chosen as a universally accepted material to provide resistance to weathering [13].

Current work is based on developing the polymeric films based on polyether-based Thermoplastic Polyurethane (TPU) along with UV absorber, AO, and HALS at various concentrations and on obtaining optimized formulation. The objective of this work is to compare the films and assess them for improving their weathering performance against UV rays in terms of UPF, gas transmission rate (GTR), yellowness index, etc.

Various classes of organic UV absorbers along with inorganic one has been explored in current work to improve the UV resistance of LTA envelope material. However, to the best of the author's knowledge, limited work on the systematic study is available on the optimization of formulation for polyether-based thermoplastic PU coatings that incorporate organic UV absorbers along with Titanium dioxide. Hence, this study is dedicated to enhance the weathering resistance of hull material for aerostats and airships using an optimized concentration of UV absorbers in the coating formulation.

2 Methodology

2.1 Materials

Polyether-based TPU (Aliphatic grade) supplied by Covestro, India, has been chosen as polymeric material for the coating matrix. White rutile TiO_2 (Mol. Mass = 79.87 g/mol) supplied by Merck Life Sciences Private Limited, Mumbai, India, has been used as an inorganic UV blocker. UV absorber, Hindered amine light stabilizer, and antioxidant purchased from CLARIANT, India, were added to the films. The molecular structures of these materials are shown in Fig. 4.

2.2 Method of Preparation of Films

TPU films of thickness 150 μ were prepared using the solvent cast method. The total loading of all materials was kept at 10%. DMF has been used as the solvent. UV absorbers were sonicated using BRANSON-3800 sonicator for about an hour well before their use to improve their dispersion in the polymer matrix. TPU is added gradually into the solution to avoid agglomeration, and the solution is then kept for magnetic stirring at 40 °C for 12 h to achieve the desired viscosity. Table 1 gives the concentration of different materials used in films named F1, F2, F3, F4, and F5 for the experiments, as shown in the table below. Films were then dried for 24 h and get tested for different properties.

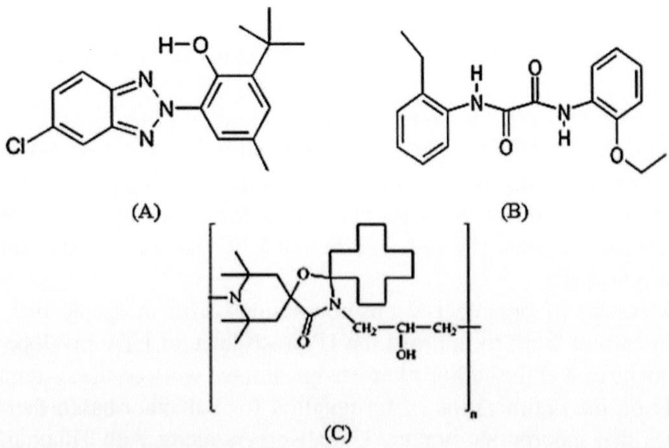

Fig. 4 Structure of **a** antioxidant, **b** benzotriazole-based UVA, and **c** hindered amine LS [14, 15]

Table 1 Composition of TPU films

Material used	F0	F1	F2	F3	F4	F5
TPU (%)	12	12	12	12	12	12
TiO_2 (% owp)	0	3	5	5	5	7
UVA (% owp)	0	3	3	2	1	1
AO (% owp)	0	3	1	2	3	1
HALS (% owp)	0	1	1	1	1	1
Total loading (% owp)	0	10	10	10	10	10

2.3 UV–VIS Spectroscopy

UV–VIS Spectroscopy of the UV absorber was done at the concentration of 0.001, while HALS and antioxidants were analyzed for UV–VIS spectra at 0.002% individually using UV-1900i UV–Vis Spectrophotometer. Also, the UV–VIS Spectra were analyzed for the solutions of the mixture of UVA, HALS, and AO at the concentration of 0.002%. Spectra of absorbance verses wavelength were obtained.

2.4 Fourier Transform Infra-Red Spectroscopy (FTIR)

FTIR spectra of the prepared film before and after weathering were obtained to identify the chemical interactions present in the structure of sample materials. This analysis is carried out using Nicolet iS50 Thermo Fisher spectrometer in attenuated total reflection (ATR) mode choosing ZnSe accessory. Number of scans taken were 64. The range of wavenumber considered is 500–4000 cm^{-1}.

2.5 Artificial Accelerated Weathering

All the prepared TPU films were exposed for 200 h in an artificial weathering instrument, i.e., Xenotest 440 ATLAS, to simulate the natural exposure at higher altitudes in the stratospheric environment as per the testing standard used is ISO 4892–2. The optical filters used were XENOCHROME 340 nm. UV exposure of TPU films was done for 200 h in accelerated conditions. The cycle gets repeated after every two hours, i.e., dry cycle for 1 h 42 min under UV irradiance 0.5 ± 0.02 Wm^{-2} nm^{-1} and wet cycle for 18 min under UV irradiance 0.51 ± 0.02 Wm^{-2} nm^{-1} with the temperature of the black panel at 65 ± 3 °C and chamber temperature at 38 ± 3 °C, RH of 50 ± 10 %.

2.6 UPF Measurement

The evaluation of resistance to UV is calculated in terms of UV Protection Factor (UPF). UPF of TPU films was determined using the LABSPHERE instrument as per the testing standard AS/NZS 4399:2017. The wavelength range was kept between 290 and 400 nm. It measures the UV light transmitted or blocked by films. The fraction of UV rays blocked in UV-A and UV-B regions is calculated along with the mean value of UPF. Testing of films was carried out before and after artificial weathering of specimens, and a UPF rating was given.

$$\mathrm{UPF} = \frac{\sum_{i=280}^{400} \mathrm{E}\lambda \times S \times \Delta\lambda}{\sum_{i=280}^{400} E\lambda \times S\lambda \times \mathrm{T}\lambda \times \Delta\lambda} \tag{1}$$

where $E\lambda$ represents relative erythemal spectral effectiveness

$S\lambda$ = solar spectral irradiance
$\Delta\lambda$ = measured wavelength interval (nm)
$T\lambda$ = average spectral transmittance of the specimen [9, 16].

2.7 Yellowness Index

SS5100 H is a dual-beam spectrophotometer used for color measurement. The wavelength range selected was 360 to 700 nm. The yellowness index is calculated as per the testing standard ASTM E-313 before and after exposing the specimens to artificial weathering. The control sample is a neat TPU film used in the testing instrument.

2.8 *Gas Barrier Measurement*

Materials to be used in LTA systems should have low helium gas permeability to improve their life. Hence, it is an important property to measure. Helium gas permeability of prepared films was evaluated using CLASSIC 216 gas permeability tester by LABTHINK as per the testing standard ASTM D1434. Results can be obtained for gas transmission rate (L/m^2/24 h). Testing gas between the two sides will form a pressure difference of 0.1 Mpa. Test area of sample is 38.48cm^2. GTR can be calculated using the following Eq. (2).

$$\mathrm{GTR} = \frac{V_c}{A \times R \times T \times Pu} \times \frac{\mathrm{d}p}{\mathrm{d}t} \tag{2}$$

where Vc represents the volume of the low-pressure side

T is the test temperature (thermodynamic temperature)
A is the effective transmission area
$\mathrm{d}p/\mathrm{d}t$ is the pressure variation on the low-pressure side per unit time after the transmission becomes stable
R is the gas constant.

2.9 *Tensile Testing of Films*

According to the test method, ASTM D5035, the INSTRON testing machine was used to determine the tensile strength of TPU films. It measures the force and elongation at break. The samples prepared were of dimensions 10 × 2 cm. The tests were performed at the extension rate of 300 mm/min and a gauge length of 5 cm. Values for breaking strength in kgf/5 cm were then calculated.

3 Results and Discussion

3.1 *UV–VIS Spectroscopy*

UV spectra of UV absorber, AO, HALS, and the mixture of UVA, AO, and HALS were obtained as shown in Fig. 5a, b, c, d, respectively. From Fig. 5a, UVA can be considered to absorb the UV-A and UV-B region, i.e., from about 300 to 400 nm.

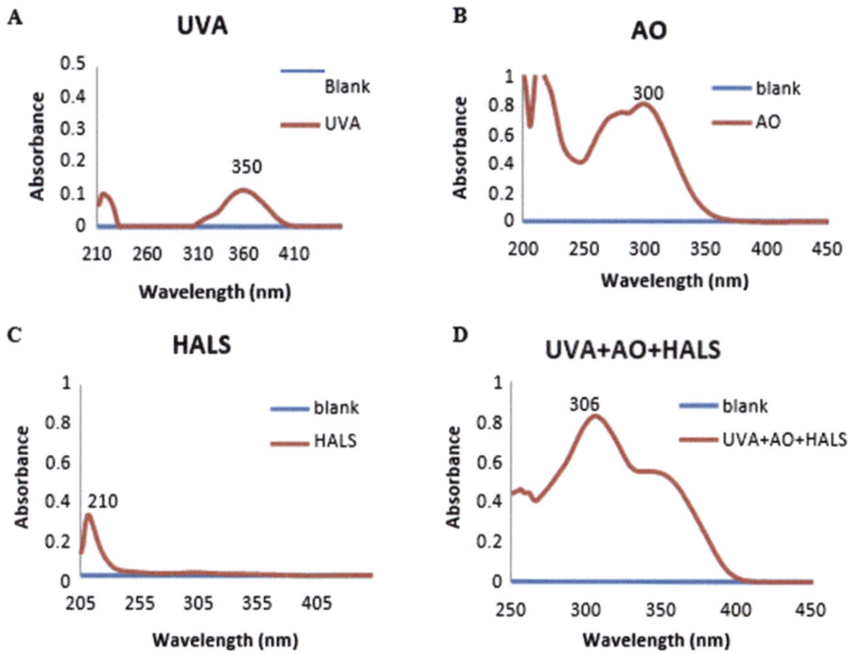

Fig. 5 UV spectra for the solution of **a** UVA, **b** AO, **c** HALS, and **d** combination of UVA + AO + HALS

UV absorber gives maximum absorbance at approximately a wavelength of 350 nm due to the presence of benzotriazole in its molecular structure. Figure 5b shows that AO gives broad UV absorption from around 250 to 350 nm with λmax at 300 nm due to the presence of the oxalanilide group, while HALS gives narrow UV spectra in the UVC region, i.e., from about 205 to 250 nm with λmax at 210 nm as shown in Fig. 5c. It is generally preferred to use UVA along with HALS as it acts as a radical scavenger. Figure 5d demonstrated that a mixture of UVA, AO, and HALS covers a broad range of UV spectrum from 250 to 400 nm with enhanced absorbance values.

So, it is suggested to use AO and HALS along with UV absorbers to synergize the effect of all for improved weathering resistance, although these organic UV absorbers have limited life and their performance got deteriorated due to photo-oxidation after prolonged irradiation of UV light; hence, these UVAs can be explored with inorganic UV absorbers.

3.2 *Fourier Transform Infra-Red Spectroscopy*

Before weathering, neat TPU film and hybrid films were analyzed for Fourier Transform Infra-Red (FTIR) spectroscopy in attenuated total reflectance (ATR) mode. The

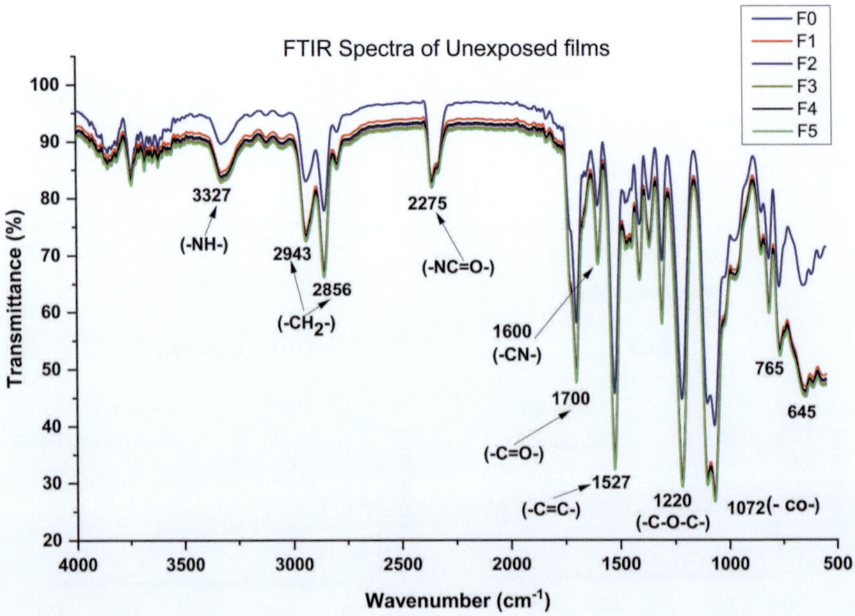

Fig. 6 FTIR Spectra of TPU films before weathering

spectrum obtained is shown in Fig. 6. The peaks at 3327 cm^{-1} in FTIR spectra show (–NH–) stretching, solid and broad spectra at 2856 and 2943 cm^{-1} show alkane (–CH_2–) stretching. The peak at 2275 cm^{-1} shows urethane (–NCO–) linkage. The bands at 1704 and 1072 cm^{-1} show dissociative (–C=O–) of the amino group in carbamic acid, and 1600 cm^{-1} shows (–CN–) linkage that confirms the presence of urethane structure in TPU film.

The peak at 1220 cm^{-1} shows (–C–O–C–) symmetric stretching of the ether linkage. The peak at 645 cm^{-1} indicated the presence of (–TiO–) metal oxide bonds in the film. FTIR spectra overlap for all hybrid TPU films; the bands at around 765 cm^{-1} show (–CH–) bending vibrations of benzotriazole benzene ring present in the structure of UV absorbers.

Neat TPU film and hybrid TPU films were compared, and it can be clearly observed from Fig. 6 that there is a decrease in transmittance % and thus ensuring good UV absorbance in the hybrid film that consists of UV absorbers along with antioxidant and light stabilizer. FTIR spectra of films after exposure to accelerated weathering are shown in Fig. 7. It can be clearly observed that the peak at 2275 cm^{-1} is completely vanished after weathering in all films. It demonstrates the breakage of urethane linkage. Also, (–NH–) stretching at 3327 cm^{-1} for neat TPU film has become almost vanished while it got broader and flattened for other hybrid TPU films, which confirms the conversion of amide linkages to urea linkages.

Similarly, the broad peaks at 2943 and 2856 cm^{-1} have become diminished for neat TPU film while it gets sharper to broader for other films. Moreover, the peak

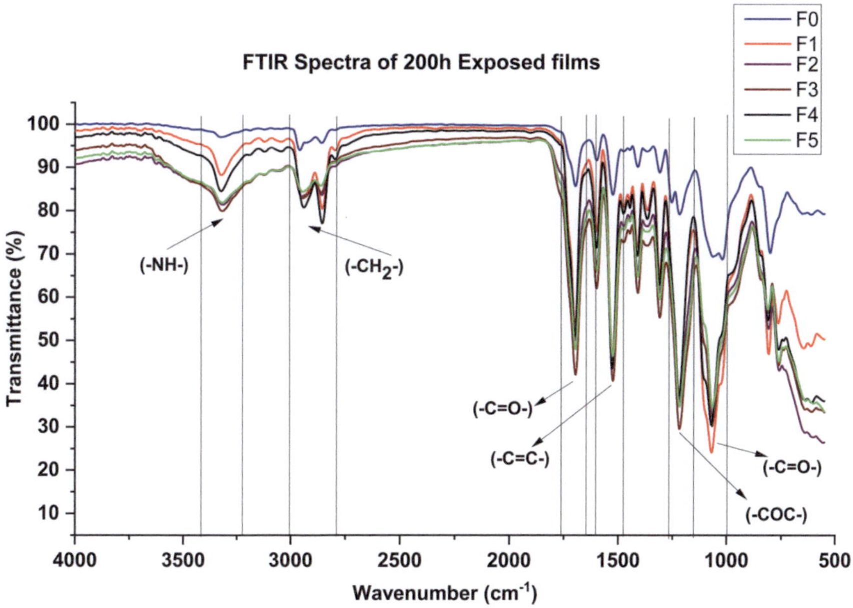

Fig. 7 FTIR Spectra of TPU films after weathering

intensity decreases at 1700, 1600, 1527, 1220, and 1072 cm^{-1} after the exposure of films to weathering. The band at 765 cm^{-1} has been retained in F1, which confirms the presence of benzotriazole in the organic UV absorber even after weathering. The band for metallic bonds at 656 cm^{-1} also got distorted, which demonstrates some molecular changes after weathering that ultimately results in a change in final properties of the TPU films such as UPF, yellowness index, gas barrier, and tensile properties.

3.3 *UPF Measurement*

UPF is measured for all TPU films that consist of TiO_2 along with organic UV absorbers and as well as for neat TPU film. From Table 2, it can be observed that neat TPU film has low resistance to UV rays as the UPF rating is 15 before weathering; although it gives good UV blocking in the UV-B region, UV-A blocking is comparatively low.

A significant surge can be seen in the case of other TPU films, i.e., about 2000. UPF rating is low, i.e., 15 for neat TPU, and it is more than 50 for other hybrid films, which show excellent UV resistance. However, UPF means the value is slightly lower for F4, but the rating again shows excellent resistance to UV. Blocking % in both the UV-A region and UV-B region can be analyzed separately.

Table 2 UPF measurement before and after accelerated weathering

Type of film	UV-A blocking (%)			UV-B blocking (%)			UPF rating		
	Before (0 h)	After (100 h)	After (200 h)	Before (0 h)	After (100 h)	After (200 h)	Before (0 h)	After (100 h)	After (200 h)
F0	74.01	91.73	93.28	97.80	99.87	99.88	15	50+	50+
F1	99.95	99.95	99.95	99.95	99.95	99.95	50+	50+	50+
F2	99.95	99.94	99.94	99.95	99.95	99.95	50+	50+	50+
F3	99.95	99.95	99.95	99.95	99.95	99.95	50+	50+	50+
F4	99.94	99.91	99.95	99.95	99.95	99.95	50+	50+	50+
F5	99.95	99.95	99.95	99.95	99.95	99.95	50+	50+	50+

Fig. 8 UPF measurement before and after weathering

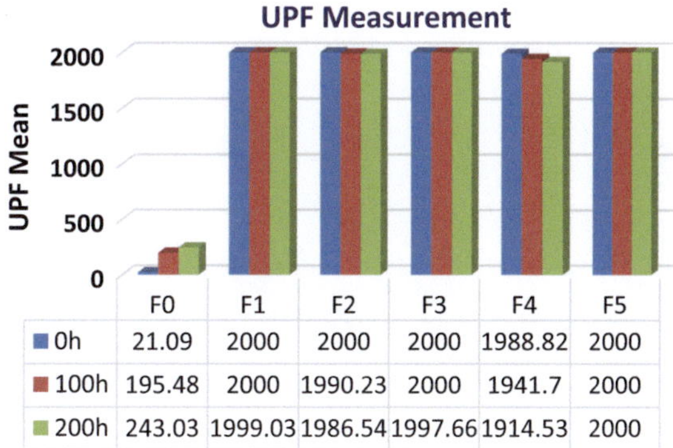

Figure 8 shows the mean values for UPF for all films. A significant surge in the UPF mean value of neat TPU film can be seen after weathering of 100 h to 200 h, i.e., 243. Most probably, the reason can be an initial reformation of bonds that leads to molecular crosslinking, as suggested by FTIR analysis. However, a slight decrease in the mean UPF value of other hybrid films can be observed after weathering, but F5 retains its original UPF value even after 200 h of weathering as the concentration of TiO_2 is relatively higher in it that helps in absorbing UV rays.

3.4 *Yellowness Index*

Neat TPU film is almost transparent and has a very low yellowness index, as given in Table 3. After incorporation of TiO_2 and other UV additives, the yellowness index increases as the amount of UV absorbers increases. Organic UV absorber is responsible for additional yellowish color to TPU films. Therefore, F1 shows the highest value of yellowness index as this film contains a high concentration of UV absorber as compared to other TPU films having a comparatively lower yellowness index.

Table 3 Yellowness index measurement before and after accelerated weathering

Type of film	Yellowness index		
	Before (0 h)	After (100 h)	After (200 h)
F0	−2.319	9.142	9.132
F1	1.599	2.598	8.689
F2	1.501	3.501	10.011
F3	1.178	2.296	19.342
F4	0.181	11.461	19.528
F5	0.804	15.398	23.331

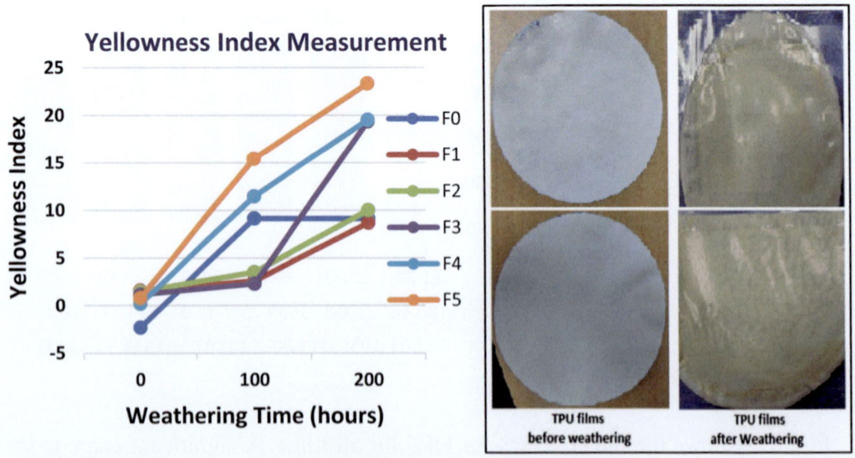

Fig. 9 Measurement of yellowness index before and after weathering

After artificial weathering of 100 h, yellowness increases in the case of all films due to photo-oxidation. However, it is maximum in the case of F5, as the amount of antioxidants is less, which contributes to the anti-yellowing properties of TPU films. The increased yellowness can be clearly observed over the increased weathering time, as represented in Fig. 9.

3.5 Gas Barrier Properties

Before weathering, the helium gas barrier property of prepared TPU films was determined, and the calculated gas transmission rate is given in Table 4 as shown. The values of GTR over weathering time are represented in Fig. 10.

After 100 h of exposure, the capability of films to retain helium has been degraded except for the neat TPU film, F3 and F5. Initial crosslinking of molecular chains over

Table 4 Measurement of GTR before and after weathering

Type of film	Helium gas permeability GTR (Ltr/m^2/day)		
	Before weathering (0 h)	After weathering (100 h)	After weathering (200 h)
F0 (Neat TPU)	2.035	2.88	1.943
F1	1.84	2.431	2.082
F2	1.832	2.468	2.289
F3	2.088	2.01	2.161
F4	2.664	2.864	3.104
F5	1.804	1.611	1.715

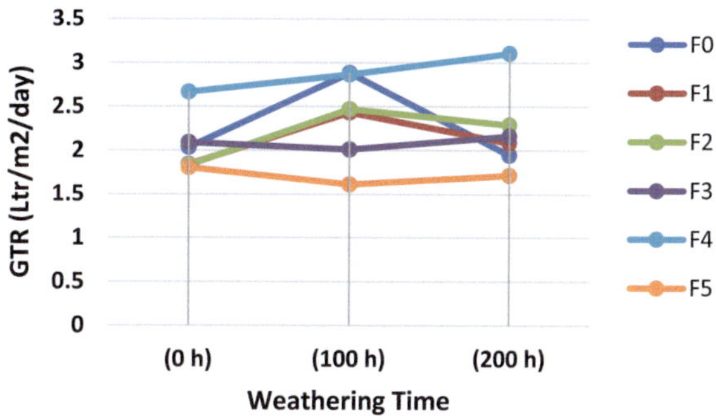

Fig. 10 Measurement of GTR before and after weathering

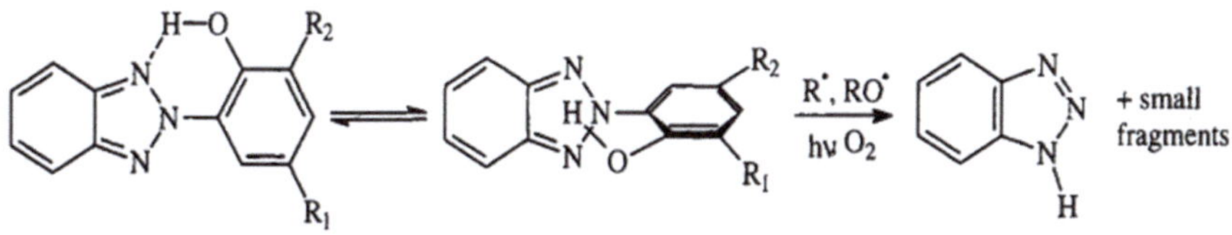

Fig. 11 Photo-oxidation of benzotriazole-based UV absorber [17]

UV exposure can be the reason. While after 200 h of exposure, Helium permeability decreases for F0, F1, and F2 while it increases for F3, F4, and F5.

The cause is obviously related to polymer degradation and photo-oxidation of UV absorbers after prolonged exposure, and the rate of degradation and photo-oxidation may vary for different TPU films. Also, it depends on the morphological structure of the materials used.

Benzotriazole-based UV absorbers are supposed to get attacked by free radicals at the phenolic ring in its excited energy state in the non-polar state rather than due to reaction at the phenolic hydroxyl group shown in Fig. 11 as suggested by other researchers also.

3.6 Tensile Testing

The results for tensile strength in terms of breaking strength of all films are given in Table 5 and Fig. 12 before and after weathering of 200 h. It can be clearly observed that there is a loss in breaking strength of all the films because of degradation of polymer and photo-oxidation. However, the extent may vary.

Breaking strength for neat TPU film is comparatively lower before weathering, and the percentage loss after weathering is also higher, i.e., approximately 18.5%. While for other hybrid films, the loss in breaking strength is lower, i.e., approximately 4–8%. F1 showed the minimum loss in breaking strength followed by F5, F4, F2, and F3. UVAs led to self-degradation over long exposure of UV in TPU films as

Table 5 Measurement of breaking strength before and after weathering

Type of film	Breaking strength (kgf/5 cm)	
	Before weathering (0 h)	After weathering (200 h)
F0	85.675	69.725
F1	108.92	104.87
F2	107.75	99.39
F3	108.55	99.87
F4	102.81	97.43
F5	110.95	106.72

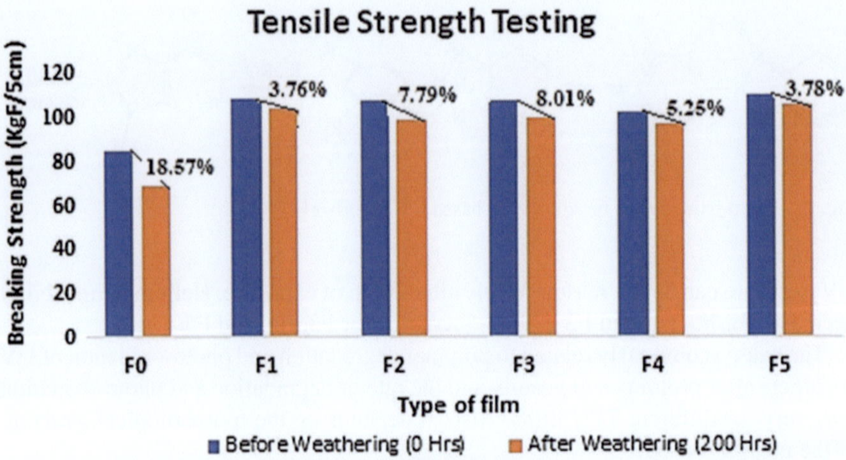

Fig. 12 Measurement of breaking strength of TPU films

represented in Fig. 11. TiO_2 gets degraded over time owing to its high photocatalytic activity. Polymer degradation also occurs as the time of exposure increases.

4 Future Scope

The trend for various properties has been analyzed with respect to weathering of up to 200 h. It can be further done for at least 500 h to precisely understand the changes that can occur during weathering. Moreover, a comparison with natural weathering can be made. Many other inorganic fillers, such as zinc oxide, cerium oxide, zirconium oxide, etc., can be explored as a substitute to TiO_2 to enhance suppressibility to photodegradation as well. A more possible combination of concentrations of different UVAs can be explored using the Design of Experiment (DOE) approach statistically to optimize parameters more accurately.

5 Conclusion

The composition of TPU films is optimized to enhance resistance against environmental degradation, particularly UV resistance. TPU films based on Titanium dioxide and organic UV absorbers, along with antioxidants and light stabilizers, significantly improve the weather resistance properties of coatings that can be utilized for developing hull material for LTAs. Helium gas retention, UPF, and tensile properties have been improved as compared to the neat TPU film.

Weathering would result in increased yellowness of all TPU films owing to photodegradation of polymer and UV additives and some minimal decrease in tensile properties, UPF, and Helium gas barrier, but neat TPU film has been degraded to a large extent in terms of all properties. As retainment of helium is very important, F5 film is supposed to improve the helium gas barrier, while other films such as F1 and F3 are also able to retain helium well. F1, which consists of inorganic UVA and organic UVA filler in almost equal amounts, was found to show the best tensile property than others, like the loss in breaking strength after exposure to weathering is very less followed by F5. UPF has been retained by all hybrid films. Moreover, F1 and F5 give the best results. F1 also showed better resistance to yellowness after weathering. All the obtained results are verified by FTIR spectra.

Hence, the concentration of inorganic and organic UV additives incorporated in TPU films is optimized. It is suggested to use AO along with organic UVAs as it prevents yellowness and HALS to obtain a synergistic effect. The combination of different concentrations of UVAs as given in F1 is preferable. It clearly demonstrated the potential of using organic UVAs along with TiO_2 for better UV resistance fulfilling the primary requisites for aerostats and airship's hull material.

Acknowledgements The authors are grateful to acknowledge the Joint Advanced Technology Centre (JATC)—DRDO lab and IIT Delhi to provide lab facilities for carrying out this research work.

References

1. Zhai H, Euler A (2005) Material challenges for lighter-than-air systems in high altitude applications. In: AIAA 5th ATIO and16th Lighter-Than-Air Sys Tech and balloon systems conferences. American institute of aeronautics and astronautics, Reston, Virigina
2. Raza W, Singh G, Kumar SB, Thakare VB. Challenges in design and development of envelope materials for inflatable systems
3. Li A, Vallabh R, Bradford PD, Seyam A-FM (2019) Textile laminates for high-altitude airship hull materials-a review
4. Dasaradhan B, Das BR, Sinha MK, Kumar K, Kishore B, Prasad NE (2018) A brief review of technology and materials for aerostat application. Asian J Text 8(1):1–12. https://doi.org/10.3923/ajt.2018.1.12
5. Adak B, Joshi M (2018) Coated or laminated textiles for aerostat and stratospheric airship

6. Stockbridge C, Ceruti A, Marzocca P (2012) Airship research and development in the areas of design, structures, dynamics and energy systems. Int J Aeronaut Space Sci 13:170–187. https://doi.org/10.5139/IJASS.2012.13.2.170
7. Wirunchit S, Apivitcholchat C, Chodjarusawad T, Koetniyom W (2018) The study of UV protection materials. AIP Conf Proc Vol 2010. https://doi.org/10.1063/1.5053200
8. Mahltig B, Böttcher H, Rauch K, Dieckmann U, Nitsche R, Fritz T (2005) Optimized UV protecting coatings by combination of organic and inorganic UV absorbers. Thin Solid Films 485:108–114. https://doi.org/10.1016/j.tsf.2005.03.056
9. Zeljko M, Ocelić Bulatović V, Špada V, Blagojević SL (2021) Environmentally friendly UV-protective polyacrylate/TiO_2 nanocoatings. Polymers 13(16):2609. https://doi.org/10.3390/polym13162609
10. Yang H, Zhu S, Pan N (2004) Studying the mechanisms of titanium dioxide as ultraviolet-blocking additive for films and fabrics by an improved scheme. J Appl Polym Sci 92:3201–3210. https://doi.org/10.1002/app.20327
11. Liu Y, Liu Y, Lin J, Tan H, Zhang C (2014) UV-protective treatment for Vectran® fibers with hybrid coatings of TiO_2/organic UV absorbers. J Adhes Sci Technol 28:1773–1782. https://doi.org/10.1080/01694243.2014.921130
12. Wang H, Wang Y, Liu D, Sun Z, Wang H (2014) Effects of additives on weather-resistance properties of polyurethane films exposed to ultraviolet radiation and ozone atmosphere. J Nanomaterials 2014 https://doi.org/10.1155/2014/487343
13. Maekawa S, Shibasaki K, Kurose T, Maeda T, Sasaki Y, Yoshino T (2008) Tear propagation of a high-performance airship envelope material. J Aircr 45(5):1546–1553. https://doi.org/10.2514/1.32264
14. Hostavin®* VSU (2011) Non discoloring UV absorber for plastic materials and coatings
15. Hostavin 3226P (2011) https://www.clariant.com/en/Solutions/Products/2014/06/17/08/05/Hostavin-3326-P
16. Standards Association of Australia/Standards New Zealand (1996) Sun protective clothing: evaluation and classification. Standards Australia
17. Gerlock JL, Tang W, Dearth MA, Korniski TJ (1995) Reaction of benzotriazole ultraviolet light absorbers with free radicals. Polym Degrad Stab 48:121–130. https://doi.org/10.1016/0141-3910(95)00027-J

Airship Turn Performance Estimated From Efficient Potential Flow Panel Method

Jesús Gonzalo, Diego Domínguez, Deibi López, and Carmen Salguero

1 Introduction

Automatic control of airships requires specific tracking techniques [1] where aerodynamic forces—lift and lateral forces—and pitch and yaw momenta are driving elements. The inherently instable pitch and yaw moment around the center of thick bodies such as airships may lead to control difficulties [2–4].

Numerical methods provide a useful approach to preliminary estimate aerodynamic performance. The finite volume techniques solve the Navier–Stokes equations in a quite reliable manner but requiring a high computer throughput. But for the particular problem of turn stabilization and control, where inviscid-flow forces are dominant, potential flow can be considered; this dramatically reduces the processing load enabling many optimization mechanisms at design phases of new airships. The proposed approach is very useful, as the optimization process usually requires analyzing a large set of parameters combination to identify the optimal location, span, and chord of the airship fins [5].

The solution for inviscid potential flow around thin airfoils can be successfully addressed analytically by perturbation models, and then extended numerically to thick objects, even in 3D, by the so-called panel method [6].

Present work exploits these features to estimate the turn performances of a stratospheric lighter-than-air model called ECOSAT, which currently flies vehicles from 2 to 30 m length as engineering models. In particular, the ECOSAT AS30 model is an autonomous electrical airship able to reach 3000 m with an endurance of 1 h.

J. Gonzalo (✉) · D. Domínguez · D. López
University of León, León, Spain
e-mail: jesus.gonzalo@unileon.es

C. Salguero
ECOSAT Airships Inc, León, Spain
e-mail: carmen.salguero-tascon@capgemini.com

D. Shukla (ed.), *Lighter Than Air Systems*, Lecture Notes in Mechanical Engineering,
https://doi.org/10.1007/978-981-19-6049-9_5

In order to de-risk stratospheric technologies, the model is equipped with all the elements necessary to stand stratospheric environment.

2 Problem Definition and Formulation

The formulation can be split into two main parts: geometry and aerodynamic model. Most of the details of the formulation are described in detail in [7], where a mesh generation mechanism was developed to model a large airship with its characteristic lengths of hull and fins.

From the geometric point of view and for the purpose of this work, a parametric definition of fins (shape and location) and control surfaces has been specifically developed, using a generic quadrilateral panel. The wake is derived from the position of these control surfaces using quadrilateral panels as well (Fig. 1). The panels of thin surfaces keep independent properties at both sides. Thick panels maintain an interior potential that is a-priori selected so that the external value provides the mean of calculating the local velocity. Additionally, the circulation in trailing edges is shed along a wake, which is modeled by thin flat panel strip.

From the aerodynamics point of view, the potential flow is considered by solving the Laplace equation by a sum of source and doubled distributions on the boundary, including the hull, the fins, the control surfaces, and the wake. Neumann boundary condition is applied to thin panels and Dirichlet boundary condition to those that are thick (assuming the internal body potential is known). In the end, the influence matrix is generated and solved iteratively.

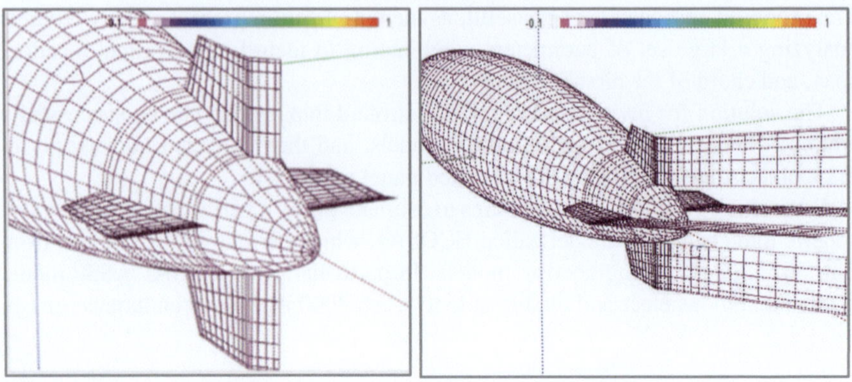

Fig. 1 Detail of the geometry of fins, control surfaces, and wake panels

3 Methodology

The novelty over the work developed in [7] is the modeling of the control surfaces and the turn rate. The rudders and elevons are installed in the four fins and defined parametrically so that geometry can be generated effortless to analyze the effect of these parameters: location, span, and chord of the airship fins.

The unperturbed stream now generates different velocities in each panel that need to be considered in the boundary condition equations that form the influence matrix.

One of the most difficult issues is to derive the panel velocity from the values of the doublets and sources associated to them. This is developed as a post processing interpolation among all the neighbors enabling the calculation of the potential derivatives, and hence, the sought induced velocity. Besides, the wake can be modified to follow local streamlines and the procedure is executed several times given the short processing time (only tenths of seconds for a 1e4 cell mesh in conventional computers).

Furthermore, the method enables the estimation of the moment derivatives with respect to the yaw/pitch rates. As it is well known, they play a fundamental role in the lateral-directional and longitudinal stability as well as turn rate performances.

4 Results

The aerodynamic prediction capabilities of the method we have developed are well illustrated by the graphs in Fig. 3. It shows the comparison between the yaw moments with respect to the vehicle center from several angles of sideslip. The test where performed in a wind tunnel using a model that included the bare hull and the fins. No rudders were included at the moment, so for the validation process only the rudder 0-deg position would be considered. Although the size of the real ECOSAT model is undisclosed, the experiment is prepared to replicate performance at length-based Reynolds about 10^6, as expected in stratospheric HAPS (Fig. 2).

An advantage of considering the resulting moments instead of the forces is that they are less dependent on the minor deviations of the pressure curves in certain areas; consequently, the matching between the experiments and the model prediction is very good. Only at intermediate angles of sideslip the potential method slightly underestimates the moment.

Furthermore, the predicted yaw coefficient is also depicted for a rudder that is totally deflected at ±25 deg, as per

$$c_n = \frac{N}{\frac{1}{2}\rho V_\infty^2 \mathrm{Vol}^{2/3} L_H} \tag{1}$$

where N is the yaw moment, ρ is the reference air density, V_∞ is the unperturbed airspeed, and Vol is the reference hull volume. In a similar way, the pitch moment

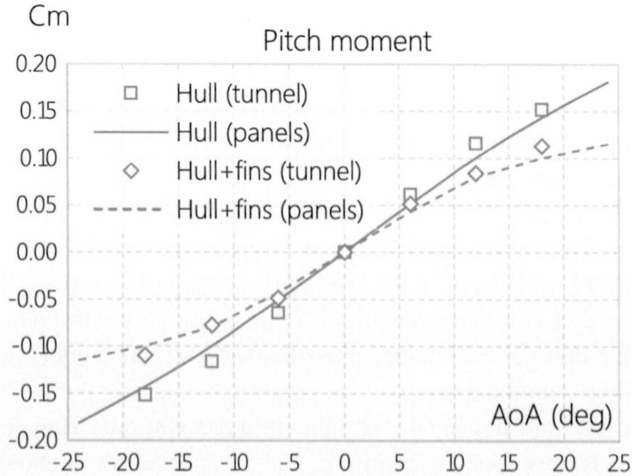

Fig. 2 Experimental an numeric pitch moment coefficients for different configurations

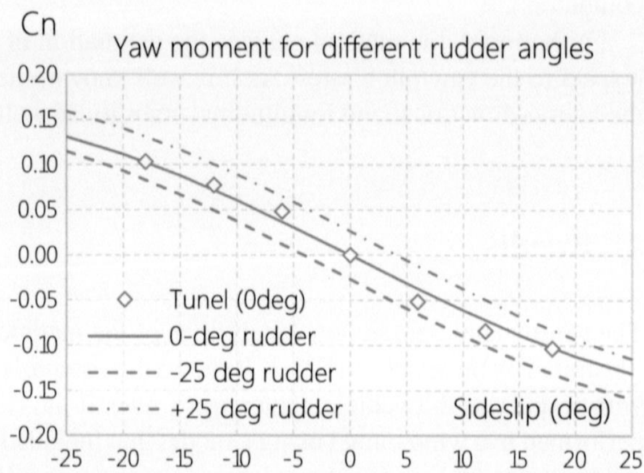

Fig. 3 Comparison of pitch and yaw moment coefficients for different rudder positions

coefficient is

$$c_m = \frac{M}{\frac{1}{2}\rho V_\infty^2 \text{Vol}^{2/3} L_H} \tag{2}$$

As expected, the yaw coefficient curve is displaced to higher or lower values depending on the rudder angle (positive or negative). Worth noting, at high angles of sideslip, the capability of the rudders to modify the yaw coefficient is not equal for positive and negative deflection angles.

The yaw moment derivative respect to the yaw rate is estimated using rotations around the center of volume to calculate the unperturbed velocity and hence to update the boundary conditions with respect to the static case. Airship turn performance is the result of combing this value with the lateral force, already implemented, and the moment induced by the yaw rate.

The definitions used in this paper to have dimensionless magnitudes are

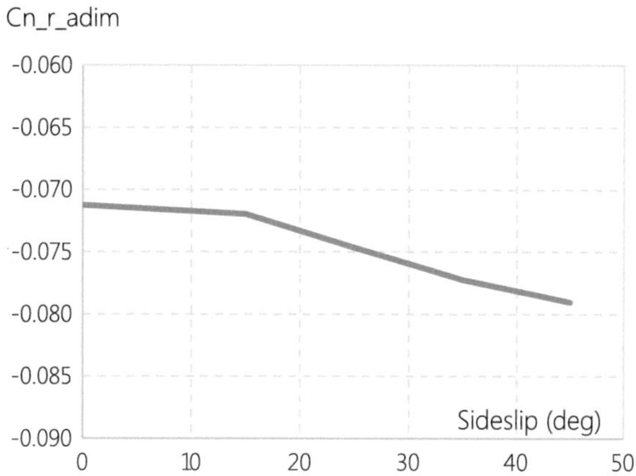

Fig. 4 Estimation of the yaw moment derivative respect to the yaw rate

$$\hat{r} = \frac{r}{L_H V_\infty} \quad (3)$$

$$c_{n\hat{r}} = \frac{\Delta c_n}{\Delta \hat{r}} \quad (4)$$

where r is the yaw rate and $\hat{r}$ its correspondent dimensionless form.

The values shown in Fig. 4 are considered reasonable when compared with available references (e.g., −0.073 for HALE bare hull [8]). Unfortunately, this value cannot be checked up to date against an experimental test.

In order to estimate the steady rotation radius, the lateral force coefficient (c_y) is deduced from the lateral force (Y). This force is less prone to be modeled with a panel method as it disregards viscosity contributions and boundary layer detachments. Figure 5 shows the underestimation of the force when compared to a 5.1×10^6 Reynolds wind tunnel test. The larger the sideslip angle the worse the results.

$$c_y = \frac{Y}{\frac{1}{2}\rho V_\infty^2 \mathrm{Vol}^{2/3}} \quad (5)$$

Finally, the radius of steady turn is extracted from numerical simulations with the above information, obtaining the results in Fig. 6. Results need to be tested in scale models but the objective of having a semi-automatic tool for the estimation of turn radius has been successfully achieved. The simulated model is quite agile as the radius is about two times the length.

5 Discussion and Conclusions

Although the design process of any airship requires highly powerful and accurate simulations and experiments, those are quite resource consuming and cannot

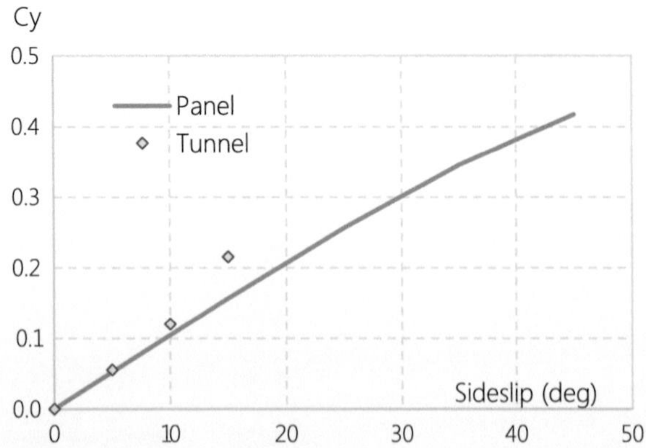

Fig. 5 Estimation of the lateral force respect to the sideslip angle

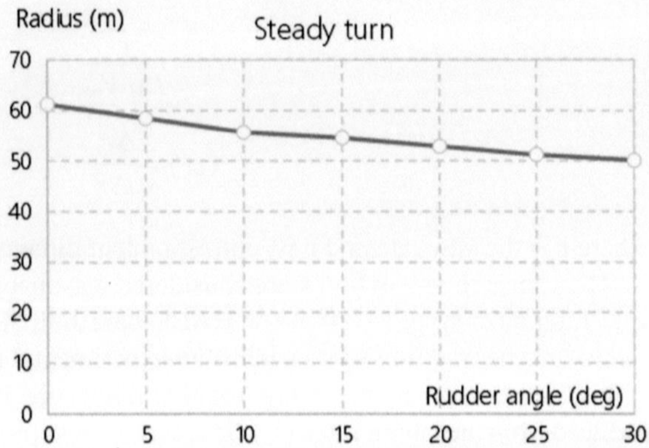

Fig. 6 Steady turn radius estimation

be massively used to explore the whole design space. That is why the idea of employing simplified—but accurate enough—methods is highly interesting during the preliminary design, when a wide variety of concepts need to be considered.

The potential method presented in this paper is capable of saving a huge amount of computational time when compared with traditional CFD, as every design configuration can be studied in a matter of a few minutes. Despite of its simplicity, it has proven to be accurate enough to provide valuable performance insights during the design process, including those related to turning performance.

Thus, yaw moment coefficients for different rudder positions can be easily calculated, as well as the yaw moment derivative respect to the yaw rate and the lateral force respect to the sideslip angle. All those data are used then to calculate the turning performance.

Preliminary results show good matching between the quick results obtained with this method and the turning performances measured by ECOSAT models in the wind tunnel. This enables a future optimization process for the final stratospheric model to have the most efficient tail definition.

References

1. Yuan J et al (2020) Trajectory tracking control for a stratospheric airship subject to constraints and unknown disturbances. IEEE Access 8:31453–31470
2. Khoury GA (2012) Airship technology. Cambridge University Press
3. Ashraf MZ, Choudhry MA (2013) Dynamic modeling of the airship with matlab using geometrical aerodynamic parameters. Aerosp Sci Technol 25:56–64
4. Carishner GE, Nicolai LM (2013) Fundamentals of aircraft and airship design: airship design and case studies. AIAA, Reston, VA
5. Suvarna S et al (2021) Optimization of fins to minimize directional instability in airships. J Aircr 59(2):317–328
6. Ashley H, Landahl M (1965) Aerodynamics of wings and bodies. Addison-Wesley
7. Gonzalo J et al (2020) On the development of a parametric aerodynamic model of a stratospheric airship. Aerosp Sci Technol 107:106316
8. Carichner GE, Nicolai LM (2013) Fundamentals of aircraft and airship design, airship design and case studies, vol 2. AIAA education series

Evaluation of High-Performance Fabric-Based Laminated Hull Material for High-Altitude Airship

Shikha Chouhan, Rishabh Tiwari, B. S. Butola, and Mangala Joshi

1 Introduction

Since long time, the world is invariably followed by a very sturdy aspiration to fly like a bird that can make a move through the air in the endless sky. In order to fulfill such kind of desire, different types of inflatables which are filled with air, as well as gases, include hydrogen, helium, oxygen, or nitrous oxide, and nitrogen is available in the market for various applications such as a balloon, inflatable castles, air-supported structure, inflatable boat, pneumatic tire, decorative inflatables, air beam, numerous air-filled swimming pool toys, and inflatable movie screen. Airships and aerostats are the most popular established examples of flying vehicles for aerospace applications with the advancement of technologies and emerging new nanomaterials.

In the recent years of advanced technologies, high-altitude lighter than air systems are subjected to 20 km above sea level considering great interest [1]. Lighter than air (LTA) systems may be categorized into two types: (a) airship (untethered system) and (b) aerostat (tethered system). The factors which are affecting the altitude are severe weather, jet flowing or wind velocity, and the Federal Aviation Administration air-traffic layer [2]. The tethered aerostat is controlled from a stationary or permanent location, and the lift is supplied exclusively by a lifting gas including hydrogen or helium which is loaded in the envelope. It needs no force for controlling the altitude or for station keeping and it is easily retrievable for maintenance of the payload. On the other side, the untethered airship is an unmanned, power-led, and free-fluttering vehicle system in which the lift is managed by a coupled effect of aerodynamics and lifting gas in order to deliver the requisite power required to obtain the target by the combination of advanced energy storage and photovoltaic system. Aerostat may be categorized by (a) hull configuration (flexible hull, semi-rigid hull, and rigid hull), (b) the vertical force-producing way (heavier than air system, lighter than air

S. Chouhan (✉) · R. Tiwari · B. S. Butola · M. Joshi
Indian Institute of Technology Delhi, Hauz Khas, New Delhi 110016, India
e-mail: shikhahchouhaniitb@gmail.com

D. Shukla (ed.), *Lighter Than Air Systems*, Lecture Notes in Mechanical Engineering,
https://doi.org/10.1007/978-981-19-6049-9_6

system, and hybrid system), and (c) capability of payload (medium-lift and heavy-lift). The flexible or non-rigid type of hull material is more proposed over others due to its lightweight, simple structure, lower cost for manufacture and maintenance, and easy fabrication [3, 4]. The LTA systems are greatly engaged in defense applications including military surveillance, discovering aerial threats, homeland defense as well as also used in weather forecasting, network monitoring, telecommunication relay, broadcasting, disaster management, multimedia, and 4G services for cellphones or wireless systems [5, 6].

A stratospheric airship can provide a value-effective replacement to earth-orbit satellites in order to achieve unique datasets associated with science and telecommunications. High-altitude airships are having some unique balanced advantages over satellites including rapid reconstitution of capabilities, low cost for accessing and launching, persistent 24/7 capability, relocate-ability, exchangeable/repairable/upgradable payloads, and long duration aloft of more than twelve months [4].

In LTA systems, one of the major structural pieces is the envelope which maintains its aerodynamic shape. When the LTA systems are subjected to various environmental circumstances, they withstand UV rays from sunlight, rain, ozone, atmospheric pressure variation, temperature variation, and pollutants thereby increasing the generation of the degradation rate of the materials used in the external layers of the envelope during their service life. The envelope should have excellent weather resistance property to cope with the terrible atmosphere in a hostile environment. Additionally, hydrogen (H_2) gas or helium (He) gases are basically used for lifting any LTA system. That is why the material should have outstanding gas barrier properties so that leakage of the gases does not take place thereby ensuring a long service life. The other necessary requirements for LTA envelop are a high strength-to-weight ratio, less moisture regains, lower creep, excellent low-temperature flexibility, good abrasion resistance, and improved hydrolysis resistance. A conventional aerostat envelope is generally constructed of polyurethane-coated polyester fabric which is laminated with a gas-holding layer. In the case of any single-layered material, it is quite difficult to fulfill all the requirements mentioned above. Therefore, multi-layered (laminated or coated) structures are generally made up to cope with severe environments in order to accomplish all other requirements [2, 6].

1.1 High-Altitude Airship System

High-altitude airship (HAA) is basically a type of non-rigid airship. It provides ever-present intelligence, the war-fighter affordable, surveillance and reconnaissance, and continuous communications connectivity during the entire battlespace. It has a diameter of up to 50 m, a length of 250 m, can carry up to 1200 kg of equipment, and is able to supply it with electrical power.

The different parts of a typical non-rigid or flexible aerostat system (also recognized as blimp) are (a) hull (outer structure) acts as an external skin made up of multilayered laminate material, (b) ballonet acts as an internal barrier made up of multilayered laminate material, (c) solar array acts for power generation made up of silicon-based solar array, (d) propeller acts for direction control made up of metallic fan equipped with motor, (e) rudder and elevator flap made up of Kevlar, PBO, or Vectran, and (f) gondola acts for payload and transportation made up of metal or wood.

2 Materials and Methods

2.1 Materials

2.1.1 Gas Barrier Layer

Two different BOPET films were collected from Jindal Poly Films and DuPont Teijin Films.

2.1.2 Protective Layer

Polyvinyl fluoride (PVF) film with was collected from Neoflex Industries LLP. The.

2.1.3 Strength Layer

The plain weave UHMWPE fabric was produced in the weaving lab (IIT, Delhi).

The plain weave Kevlar fabric was procured from Pee Cee Textile Stores, Kanpur.

2.1.4 Adhesive

PU synthetic adhesive and its crosslinker were collected from Pidilite Industries. Both the materials were stored in the refrigerator at a low temperature in sealed condition.

2.1.5 Other Materials

Thermoplastic polyurethane (TPU) chips, acetone, DMF, UV additives including UV absorber and hindered amine light stabilizer (HALS) were procured from Sigma Aldrich.

2.2 Methods

2.2.1 Substrate Preparation

Spectra fabric was produced with a plain weave design. The fabric was treated with 3 g/l non-ionic detergents (Lisapol N) at a temperature of 80 °C for a duration period of 1.5 h with a 1:30 material to liquor ratio and 7.6 pH in order to remove the residual spin finish, hydrophobic foreign materials, and other additives. After that, the fabric was thoroughly rinsed off with cold water and dried at 100 °C temperature for a period of 30 min. The fabric was then cut into small specimens of the required size. BOPET films and PVF films were also cut into samples of A4 size.

2.2.2 Plasma Treatment

Spectra fabric, PVF films, and BOPET films were cut into 12×15 cm^2 sizes. The plasma machine can generate a 10 kHz plasma frequency with a power supply of 17 kV voltage atmosphere. The machine has 5 strip types of electrodes that are run without moving the sample in the presence of atmospheric gas including oxygen, nitrogen, and helium gas. The samples were mounted on the machine at taught conditions with the help of a lead cloth in such a manner so that sample does not come in contact with electrodes on both surfaces. The dwell period of plasma was varied from 1 to 4 min in the presence of helium and oxygen gas. After each cycle, the machine was provided a break of 30 min.

2.2.3 Preparation of PU Solution

The required amount of neat PU chips (5–10% with respect to solvent) was taken in the aluminum foil to dry in the vacuum oven at a temperature of 80 °C for a duration of 3 h in order to remove the moisture properly. The soaking of the chips in the solvent (DMF) was carried out overnight (at least 12 h) at room temperature in order to cause swelling of the chips properly with the help of magnetic bits in a stationary condition within a 250 ml beaker in a sealed condition with paraffin films. The swollen material was treated with gradient heating starting from 50 °C temperature for 1 h, 60 °C for one hour, 70 °C for 1 h, and 75 °C for an hour by using an oil bath hot plate. After 3 h of completion of gradient heating, magnetic stirring was carried out in the speed range of 30 rpm to 90 rpm when PU chips were almost dissolved in the solvent.

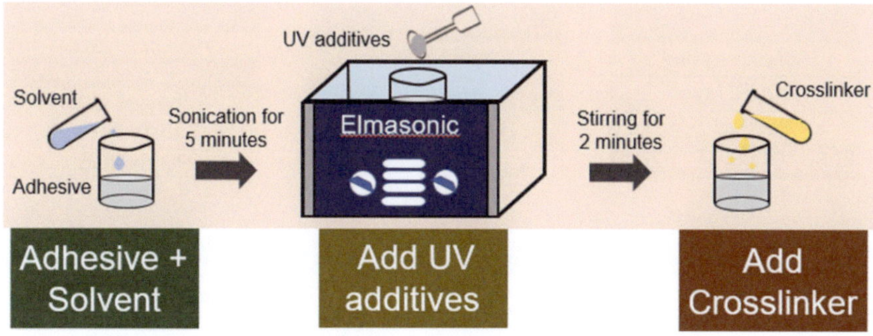

Fig. 1 Schematic process flow of adhesive formulation

2.2.4 Adhesive Formulation

In order to reduce the viscosity of the adhesive (PU-based), the required amount (10–15%) of solvent (Acetone) was added into the resin to improve the ease of handling. After that, UV absorber, HALS, and antioxidants were properly mixed into the resin followed by sonication at room temperature for 15–20 min in order to make it bubble free and homogenized. Then crosslinker (varied in the range of 1–3%) was added to the resin followed by continuous sonication for a period of 10–12 min. The sonicated mixture of the formulated adhesive was directly applied to the plasma-treated substrates as soon as possible in order to achieve the best performance out of it. The paraffin film was tightly covered on top of the beaker during sonication to stop evaporation of the solvent from the resin. The schematic flow diagram of the adhesive formulation is shown in Fig. 1.

2.2.5 Preparation of Laminates

The freshly prepared polyurethane synthetic adhesive formulation was taken to join three individual different substrates. The adhesive was applied to the fabric surface by using a knife for controlling the thickness of the adhesive layer. After that, BOPET film was joined as soon as possible on top of the adhesive layer. The knife was further used for removing the bubbles inside two layers. Then the sample was immediately cured by evaporating the solvent with the help of a compression molding machine with different curing pressure and temperature. The chosen parameters (curing time) of the compression molding machine are given below.

- Pre-heat time: 1 min
- Full-pressing time: 2 min
- Cooling time: 1 min
- Temperature: 120
- Pressure: 125.

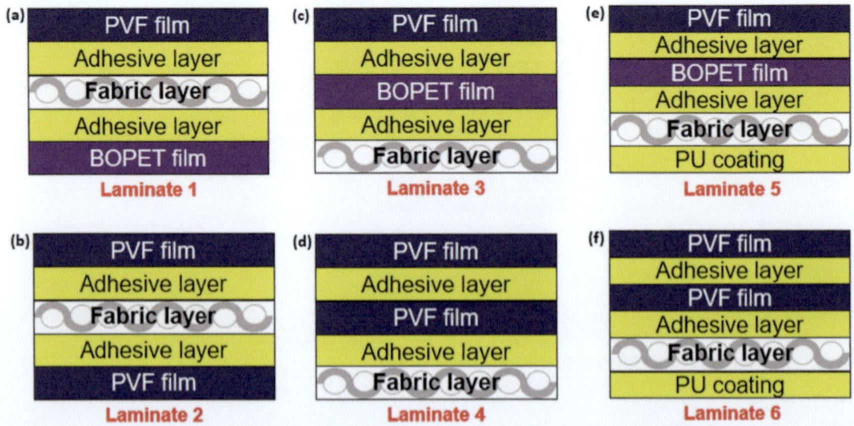

Fig. 2 Schematic diagram of different types of laminates: **a** Laminate 1, **b** Laminate 2, **c** Laminate 3, **d** Laminate 4, **e** Laminate 5 and **f** Laminate 6

In the machine, two Teflon (high temperature resistant) sheets were used to cover the sample within two stainless steel plates (15×12 cm^2), in which the sample was placed during the process. After that adhesive was cross-linked permanently between two layers. After joining of BOPET film with fabric, the adhesive was further applied on the BOPET film or another side of the fabric (unjointed side) and then PVF film was joined on top of it. Finally, the sample was further cured to get the laminates with the help of a compression molding machine with the same parameters as mentioned above. The different configurations of laminates are shown below in Fig. 2.

2.2.6 Exposure Under Accelerated Artificial Weathering

All the prepared samples (uncoated fabrics, adhesive-coated fabrics, and laminated fabrics) were subjected to exposure in a Q-SUN Xenon Test Chamber (Model: Xe-2) machine. As per ISO 4892-2 standard, the prepared samples were controlled in the artificial atmospheric conditions (temperature, humidity, and/or water spray) and exposed with a daylight filter (340 nm wavelength) under a filtered Xenon-arc light. The exposed condition had a repetitive cycle of two hours divided into a couple of steps, which were:

- Dry cycle: It is a UV exposure cycle only without water spray. Irradiance = 0.51 W/m^2.nm, time = 1 h 42 min, black panel temperature = 65 °C, chamber temperature = 38 °C, humidity = 50%.
- Wet cycle: It is a UV exposure cycle with artificial rain or water spray. Irradiance = 0.51 W/m^2. nm, time = 18 min, water spray rate = 100 ml/min.

All the samples were exposed for a duration from 100 to 400 h in which above-mentioned exposure cycles were repeated every 2 h.

3 Characterizations and Testing

3.1 Wide Angle X-ray Diffraction (XRD)

A wide angle X-ray diffraction instrument (Philips X'Pert Pro Panalytics) was used to study the crystallinity of different polymeric (BOPET film, PVF film) films. The study also helps to calculate the d-spacing value. The technique was carried out using Cu Kα radiation ($\lambda = 1.54$ nm) with 40 kV voltage and 30 mA current generation. The 2θ value was varied in the range of 5° to 40° with a scanning rate of 2°/min in the case of BOPET films.

3.2 Differential Scanning Calorimetry (DSC)

The differential scanning calorimetry of BOPET film was carried out with a TA instrument DSC machine. The weight of the material for the DSC technique was 1.2 mg. The temperature during the operation was varied from 40 to 300 °C with heating and cooling rates of 10 °C/min.

3.3 Drop Shape Analyzer (DSA)

The contact angle measurement of untreated and plasma-treated polymeric films was carried out by DSA (Drop Shape Analyzer 100) machine. A 2×1 cm^2 size sample was placed on the glass slide, and the sample was tightly bonded to it by sellotape to create a smooth surface. Each sample was measured at five different places and the average contact angle data was noted.

3.4 Atomic Force Microscopy (AFM)

The surface appearance and roughness of plasma-treated polymeric films were characterized by AFM (Bruker, Singapore) instrument in tapping mode. A 1×1 cm^2 size sample was placed on the glass slide, and the scanning area of the sample was 5×5 μm^2.

3.5 T-Peel Test

Peel strength of different two-layered laminates was carried out with a Tinius Olsen Universal Testing machine as per ASTM D1876-08 standard in order to study the adhesion behavior between film-to-film as well as fabric-to-film. As shown in Fig. 3, the laminated samples were cut with a size of 229 × 25 mm^2 along with 76 mm pre-peeled parts to hold in the jaws. The cross-head speed of the machine was 300 mm/min. The maximum load required along with extension at that load was noted in separating the layers.

3.6 Tensile Test

Breaking strength and breaking extension of different fabrics and different laminates were carried out with a tensile testing machine (Tinius Olsen Universal Tensile Tester) to observe both the maximum load applied to the sample in stretching it to rupture and the corresponding elongation of the sample in the following method of IS: 7016 (Part II). The size of the small laminated sample was 20 × 5 cm^2. The gauge length of the specimen was 75 mm, and the tests were conducted at a cross-head speed of 300 mm/min. The pre-tension on the fabrics required for the testing was 5 N when the average fabric weight was between 200 g/m^2 and 500 g/m^2. The maximum breaking load needed to break the sample was recorded along with breaking extension at that load. Five specimens were made with the length parallel to the warp threads and another five specimens were made with the length parallel to the weft threads. The average data of five samples was noted down. The tensile test was carried out at 27 °C

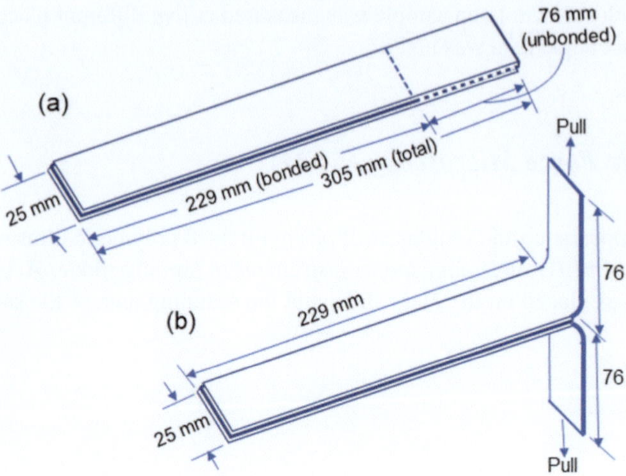

Fig. 3 Sample specimen for T-peel test during **a** preparation and **b** measuring

temperature and 65% relative humidity after conditioning the samples for 48 h prior to testing.

3.7 *Tearing Test*

The tearing strength of all the prepared samples was measured on a Tinius Olsen Universal Testing Machine as per the following method 5100 of Federal requirement 191A "Textile Test Methods." As shown in Fig. 4, the sample size for the tearing test was 6 × 4 $inch^2$ with a 1.25-inch-wide slit made with a scissor or razor blade across the exact middle (center) of the sample. The cross-head speed during the tearing test was 300 mm/min. The size of both the clamps (upper jaw and bottom jaw) was 1 inch × 1 inch. At the start of the test, the gap between clamp jaws must be 3 inches. The slit in the sample was placed in such a way so that it maintained an equal distance from each clamp. The length of the clamps was 2 inches during the testing.

3.8 *Helium Gas Permeability Test*

The gas permeability test against helium gas was carried out using N500 gas permeation analyzer (GBPI Packaging Instruments Co. Ltd.) with a constant pressure difference of 100 kPa, a temperature of 23 °C, and 65% relative humidity for PVF film, BOPET films, and laminates by ASTM D1434-82 method. The sample size for the experiment was 11 cm in diameter. The gas transmission rate (GTR) was measured in a controlled way by software installed in the computer with the change of pressure.

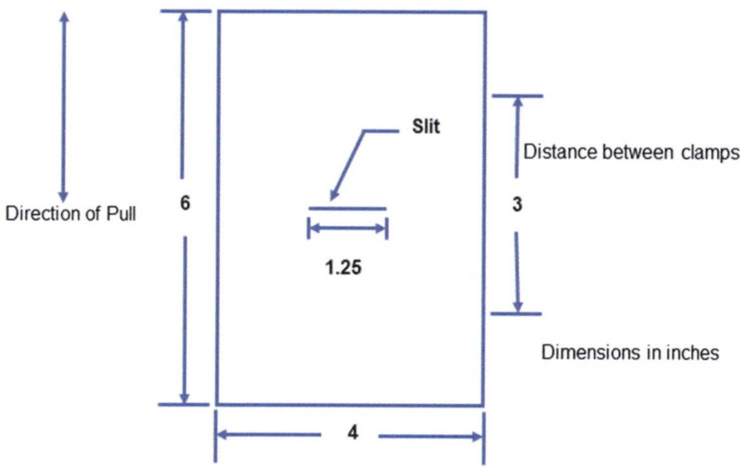

Fig. 4 Sample specimen for cut slit tearing test

3.9 UV Protection Factor (UPF)

The UPF values of adhesive-coated fabrics were evaluated through a UV transmission analyzer (Labsphere 2000F) as per AATCC 183:2004 standard. The data of UPF values was noted down in the UV wavelength range of 290–400 nm. Each sample was scanned at five different places and the average UPF data was reported.

4 Results and Discussion

4.1 XRD Study

The XRD study of both BOPET films from Jindal Poly films and DuPont Teijin Films is shown in Fig. 5. A sharp peak at 25.980 and 25.920 was observed in the case of Jindal BOPET film and DuPont BOPET film, respectively. The BOPET film from Jindal Polyfilms has shown a little higher crystallinity than that from DuPont Teijin Films, which is reflected in the higher peak intensity for J-BOPET (Fig. 5).

4.2 DSC Study

The DSC study of both BOPET films from DuPont Teijin Films and Jindal Polyfilms was carried out. The melting points of the two films are 255.9 °C and 254.4 °C, respectively (as can be seen in Fig. 6). The melting enthalpy of the films is 22.9 J/g

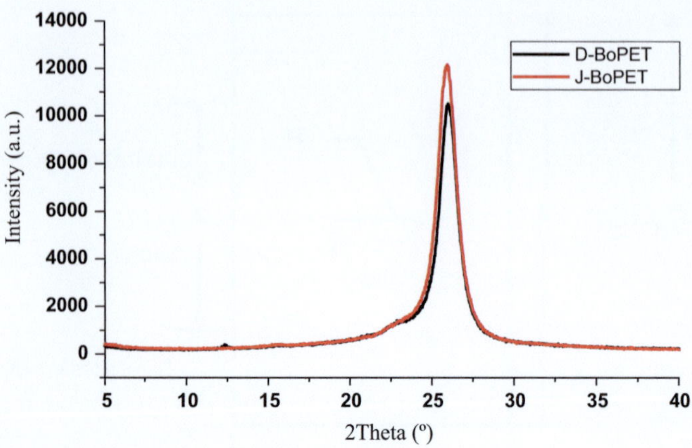

Fig. 5 XRD study of BOPET films (made from Jindal and DuPont)

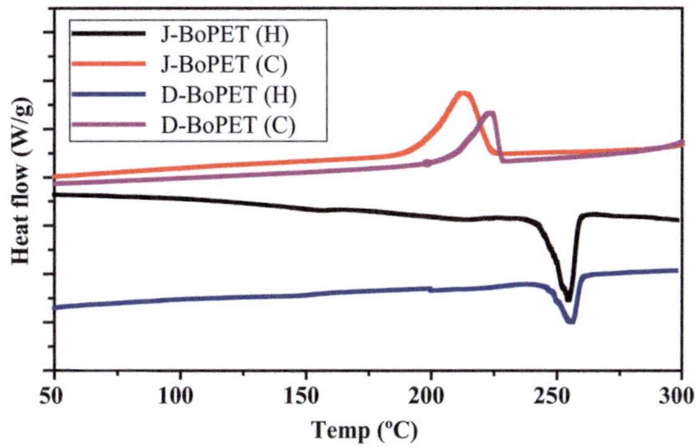

Fig. 6 DSC thermogram of DuPont BOPET film and Jindal BOPET film

and 44.2 J/g, respectively. The calculated crystallinity of the films is 16.4% and 31.6%, respectively.

In both XRD and DSC studies, the BOPET film made from Jindal Polyfilms has shown higher crystallinity than that from DuPont Teijin Films, which may result in higher gas barrier property for Jindal BOPET films.

4.3 Helium Gas Permeability

The gas permeability against helium was carried out for PVF film and BOPET film. Helium gas permeability of BOPET films is 1350 cc/m^2/day and 1662 cc/m^2/day for Jindal Poly films and DuPont Teijin Films, respectively, measuring at a fixed temperature of 230C. The permeability of PVF film is 1720 cc/m2/day at 230C temperature (Table 1).

4.4 Adhesion Studies

Contact Angle Analysis

The contact angle measurement of two different films (BOPET film and PVF film) was carried out for untreated and plasma-treated samples. The films were subjected

Table 1 Helium gas permeability value of different samples

Material/sample	Helium gas permeability (cc/m^2/day)
PVF film	1720
Jindal BOPET film	1350
DuPont BOPET film	1662

to different plasma doses in the presence of various plasma gases with a duration of varied exposure times.

As shown in Fig. 7a, the contact angle of BOPET film was decreased from 76° to 45° in the presence of helium gas and decreased from 76° to 33° in case of 70:30 He/O_2 gas and decreased up to 24° in case of 60:40 He/N_2 gas. As shown in Fig. 7b, the contact angle of PVF film was reduced from 70° to 48° in the presence of helium gas and reduced from 70° to 20° in case of 70:30 He/O_2 gas and reduced up to 27° in case of 60:40 He/N_2 gas.

Atmospheric nitrogen and oxygen actually help in forming hydrophilic groups on the surface of the substrate. Helium gas helps in cleaning the surface and also assists in breaking the ester bonds of BOPET film. It can be seen that the contact angle was significantly reduced due to the etching on both the surface of the film followed by the attachment of the polar groups on the surface. An appropriate amount (plasma dose) of oxygen gas can provide polar groups including hydroxyl (–OH) and carboxyl (–COOH) groups. The optimum amount of nitrogen gas can generate hydrophilic groups like amide (–NH2). Optimized plasma dose was observed at around 40 J/cm^2.

As shown in Fig. 8a, the contact angle of BOPET film and PVF film was reduced up to 240 and 200, respectively, for a duration of 8 min plasma exposure time. After 8 min of exposure, the contact angle increased thereby reducing surface roughness. As shown in Fig. 8b, the surface energy of BOPET film and PVF film was increased from 42 mN/m to 63 mN/m and 28 mN/m to 43 mN/m, respectively, for 8 min plasma exposure time.

However, the effect of plasma ceases to be visible with time which reveals that its effect is not permanent actually. It can be seen from the above figure that the contact angle of the substrate started to further increase after 8 min of plasma exposure. This has happened because attached polar groups were effective for a very short time period with a bulk polymer interface. The water drops shape images of untreated films and optimized plasma-treated films (PVF film with He/O_2 plasma in the ratio

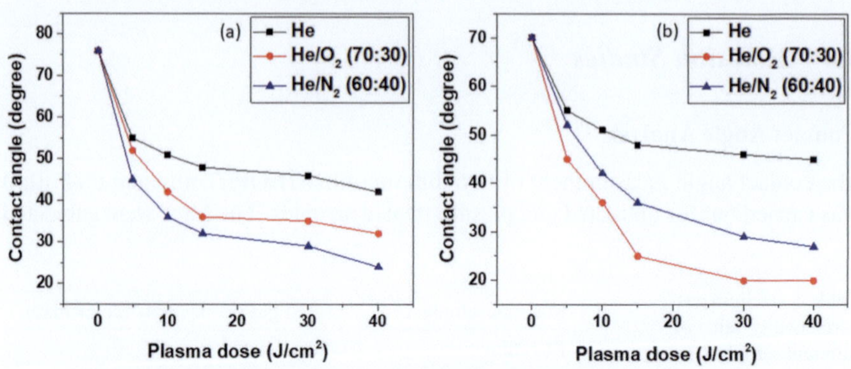

Fig. 7 Contact angle of **a** BOPET film and **b** PVF film with different plasma doses with different plasma gases (Black line indicates He gas; red line indicates the mixture of He/O_2 gas in the ratio of 70:30, and the blue line indicates the mixture of He/N_2 gas in the ratio of 60:40)

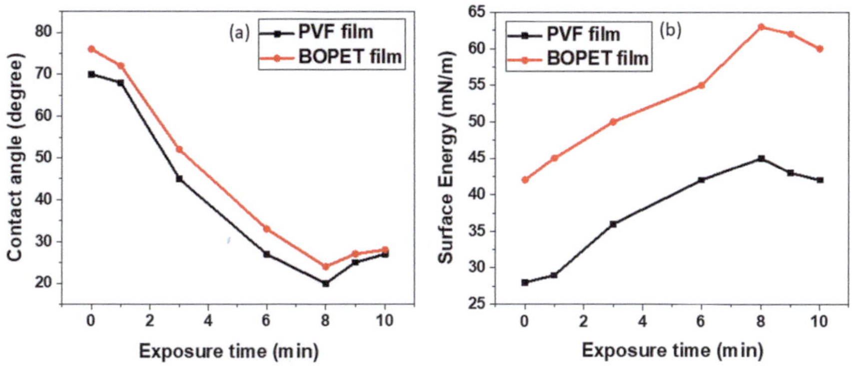

Fig. 8 **a** Contact angle and **b** Surface energy of both BOPET film and film with different plasma exposure times

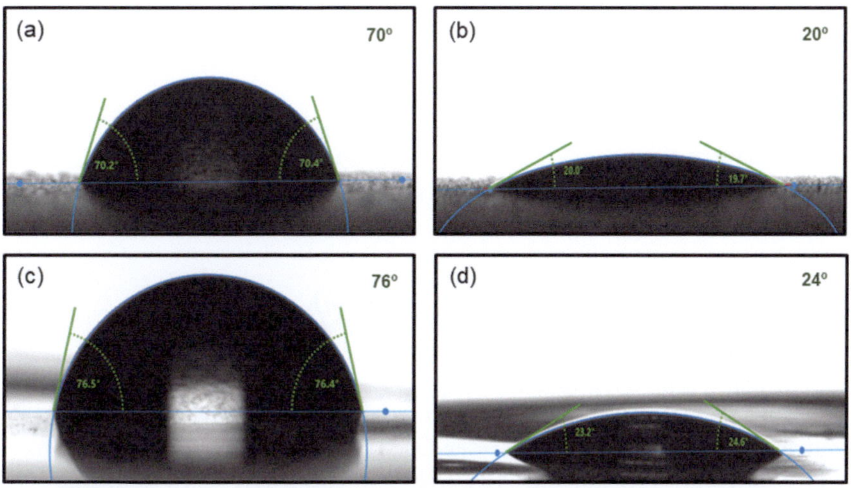

Fig. 9 Water drop shape images of **a** untreated PVF film, **b** plasma (with 70:30 He/O_2 gas)-treated PVF film, **c** untreated BOPET film and **d** plasma (with 60:40 He/N_2 gas)-treated BOPET film

of 70:30 and BOPET film with He/N2 plasma in the ratio of 60:40) were shown in Fig. 9. The optimized plasma effect was carried out at 40 J/cm^2 with a duration of 8 min atmospheric plasma.

4.5 AFM Analysis

The AFM images of two different films (BOPET film and PVF film) were taken for untreated and optimized plasma-treated samples. The AFM images indicate that the

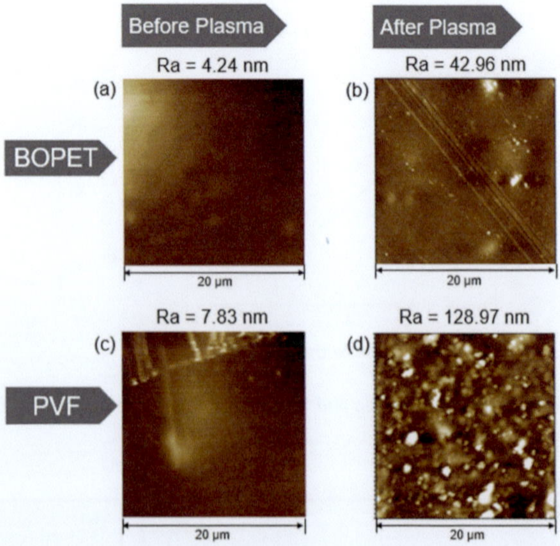

Fig. 10 Figure 4.6 AFM images of **a** untreated BOPET film, **b** plasma (with 60:40 He/N2 gas)-treated BOPET film, **c** untreated PVF film, and **d** plasma (with 70:30 He/O_2 gas)-treated PVF film

surface roughness of the films was enhanced after plasma treatment. The roughness value of BOPET film increased from 4.24 nm to 42.96 nm and it was increased from 7.83 nm to 128.97 nm in the case of PVF film (Fig. 10).

4.6 *Effect of Crosslinker on Peel Strength*

Peel strength of different double-layered joints was measured before and after plasma treatment with varied concentrations of crosslinker. After plasma treatment, the peel strength of different joints was little bit enhanced with the use of a 2.5–3% crosslinker.

Five different concentrations (1.5, 2, 2.5, 3, 3.5%) were taken to optimize the crosslinker content during adhesive preparation. As shown in Fig. 11a, the peel strength of the PVF-Kevlar joint increased from 8.4 N/cm to 15.7 N/cm before plasma and increased from 7.8 N/cm to 18.3 N/cm after plasma with 3% crosslinker. It can be seen from Fig. 11b that peel strength of the BOPET-Kevlar joint increased from 8.7 N/cm to 16 N/cm before plasma and increased from 9.9 N/cm to 18.9 N/cm after plasma with 3% crosslinker.

As can be seen in Fig. 11c, peel strength of PVF-Spectra joint increased from 6.6 N/cm to 13.6 N/cm before plasma and increased from 6.9 N/cm to 15.6 N/cm after plasma with 3% crosslinker. It can be seen from Fig. 11d, peel strength of BOPET-Spectra joint increased from 7.4 N/cm to 13.8 N/cm before plasma and increased from 8.9 N/cm to 16.1 N/cm after plasma with 3% crosslinker.

Figure 11e shows that the peel strength of the PVF-BOPET joint increased from 14.2 N/cm to 19.5 N/cm before plasma and increased from 12.8 N/cm to 24.3 N/cm after plasma with 2.5% crosslinker. Figure 11f shows that the peel strength of the

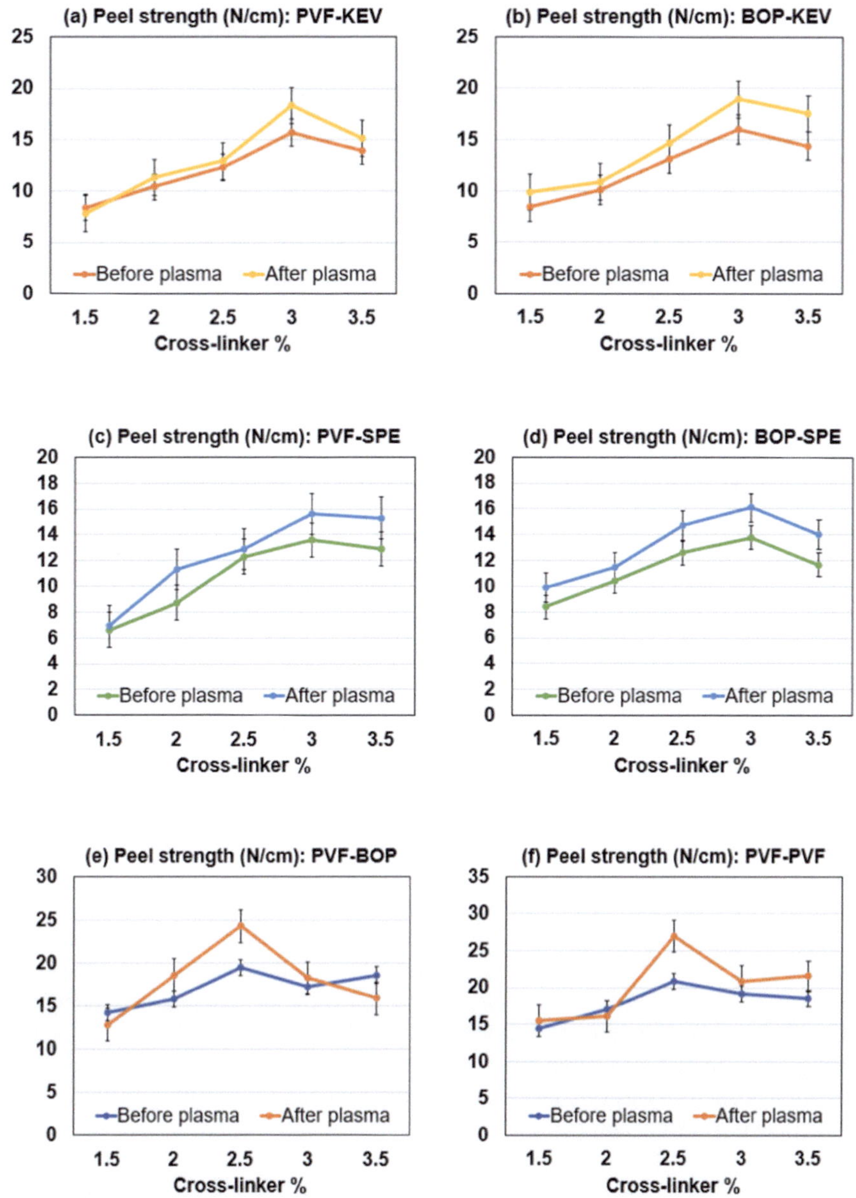

Fig. 11 Peel strength with crosslinker content of **a** PVF-Kevlar joint, **b** BOPET-Kevlar joint, **c** PVF-Spectra joint, **d** BOPET-Spectra joint, **e** PVF-BOPET joint, and **f** PVF-PVF joint, showing data before and after plasma treatment

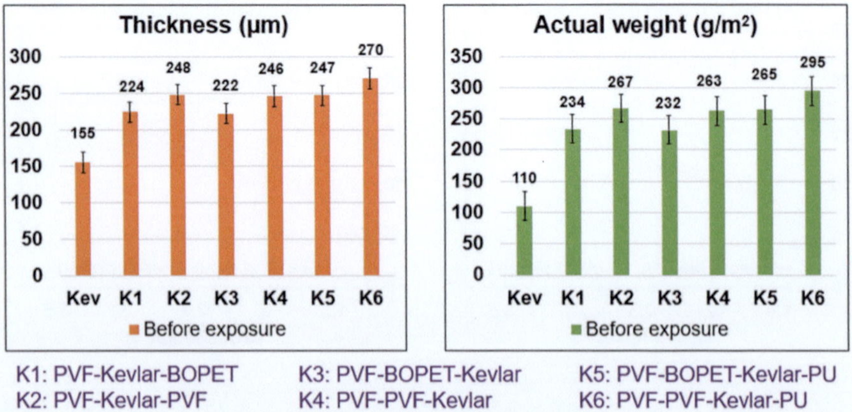

Fig. 12 Thickness and weight of different laminates made from Kevlar fabric and different films

PVF-PVF joint increased from 14.4 N/cm to 20.9 N/cm before plasma and increased from 15.5 to 27 after plasma with a 2.5% crosslinker.

5 Study on Adhesive-Coated and Laminated Fabrics

5.1 Thickness and Weight of the Laminates

The thickness of neat Kevlar fabric was 155 µm before UV exposure as shown in Fig. 12 The thickness of different configurations (K1, K2, K3, K4, K5, and K6) of laminates was varied from 224 µm to 270 µm.

The actual aerial density or weight per unit area of Kevlar-based laminates was varied from 234 to 295 g/m^2. As can be seen from Fig. 13, the thickness of neat Spectra fabric was 192 µm before UV exposure. The thickness of different configurations (S1, S2, S3, S4) of laminates was varied from 250 to 275 µm. The actual aerial density of Spectra-based laminates was varied from 272 to 302 g/m^2. However, the deviation in the data was very less as shown in Figs. 12 and 13 resulting in uniformity in the thickness of the laminates.

5.2 Effect of UV Additive on Peel Strength

Peel strength of various double-layered joints was evaluated by using without and with UV additives (having UV absorbers, HALS, and antioxidant). Peel strength of all the joints significantly improved after using UV additives. Figure 14 shows that the peel strength of the PVF-Kevlar joint increased from 8.9 to 14.7 N/cm^2, whereas

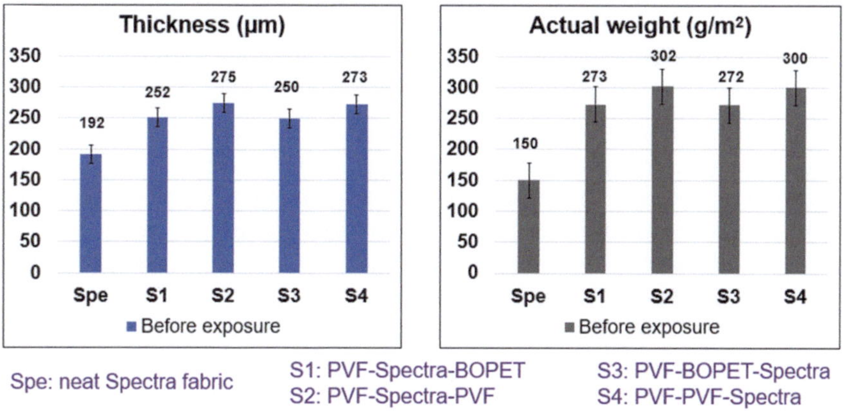

Fig. 13 Thickness and weight of different laminates made from Spectra fabric and different films

it was increased from 10.4 to 15 N/cm^2 in the case of the BOPET-Kevlar joint. Peel strength of PVF-Spectra joint increased from 10.2 to 15.6 N/cm^2, whereas it was increased from 12.2 to 16.2 N/cm^2 in the case of BOPET-Spectra joint.

Peel strength of PVF-BOPET joint (i.e., joint between two different films) increased from 14.4 to 17.1 N/cm^2, whereas it increased from 17.1 to 19.5 N/cm^2 in the case of PVF-PVF joint (i.e., joint between two similar kinds of films). The peel strength of the film-film joint was greater than that of the film-fabric joint. Among all the joints, peel strength of the PVF-PVF joint was maximum due to the joint of similar kind of films.

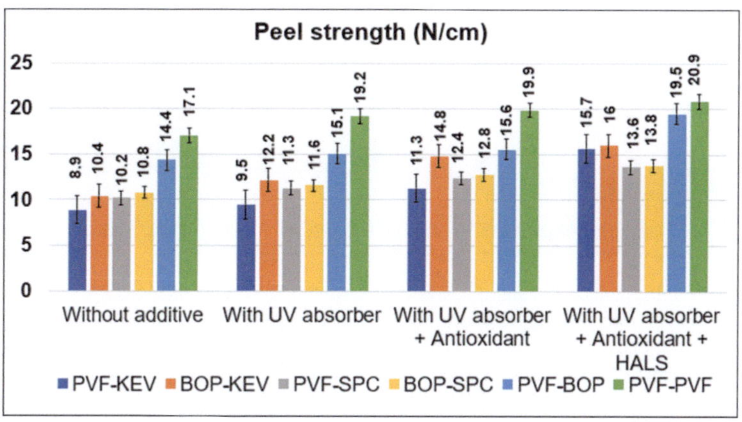

Fig. 14 Peel strength of different double-layered joints (PVF-Kevlar joint, BOPET-Kevlar joint, PVF-Spectra joint, BOPET-Spectra joint, PVF-BOPET joint, and PVF-PVF joint) with different UV additive combinations

5.3 *Effect of UV Additive and Accelerated Weathering on UPF Value*

UV protection factor (UPF) is a measure of the efficacy of protection against UV light or sunlight coming to the earth. The UPF of any textile material is a function of chemical characteristics, construction of the fabric, moisture content of the fabric, porosity, thickness, extension of the fabric, color, and the finishing given to the fabric. A higher UPF value stands for higher UV absorbency thereby resulting in greater protection from degradation by UV light.

Figure 15 shows the change in UPF values of uncoated and adhesive-coated Kevlar fabrics made from different UV additive combinations with 100 h and 200 h UV exposure. The UPF of uncoated and unexposed Kevlar fabric was 257 which was reduced to 96 after 200 h of UV exposure thereby resulting in higher degradation by UV radiation. The UPF of adhesive (without additive)-coated and unexposed Kevlar fabric was 1704 which was reduced to 933 after 200 h of UV exposure, whereas the UPF of adhesive (mixed with all UV additives)-coated and exposed Kevlar fabric was 1932 even after 200 h of UV exposure thereby resulting in higher UV absorbency and grater UV protection.

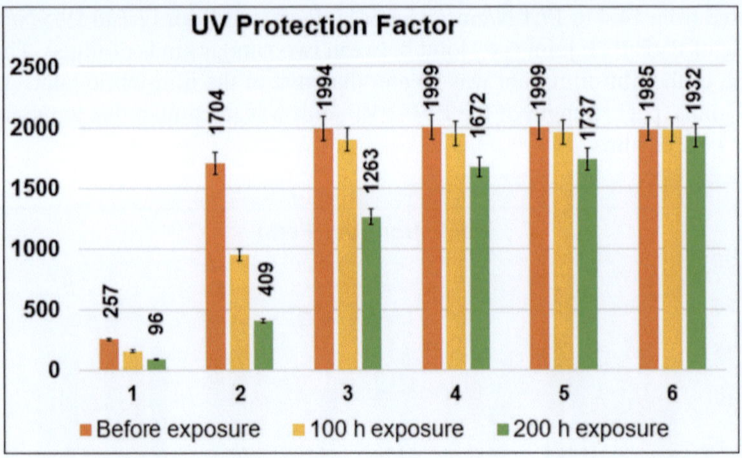

Fig. 15 UPF value of uncoated and adhesive-coated Kevlar fabric with different UV additive combinations [1. Neat Kevlar fabric, 2. Adhesive-coated Kevlar fabric without additive, 3. Adhesive-coated Kevlar fabric with HALS, 4. Adhesive-coated Kevlar fabric with UV absorber, 5. Adhesive-coated Kevlar fabric with UV absorber and HALS, and 6. Adhesive-coated Kevlar fabric with UV absorber and HALS and antioxidant], showing data before and after UV exposure

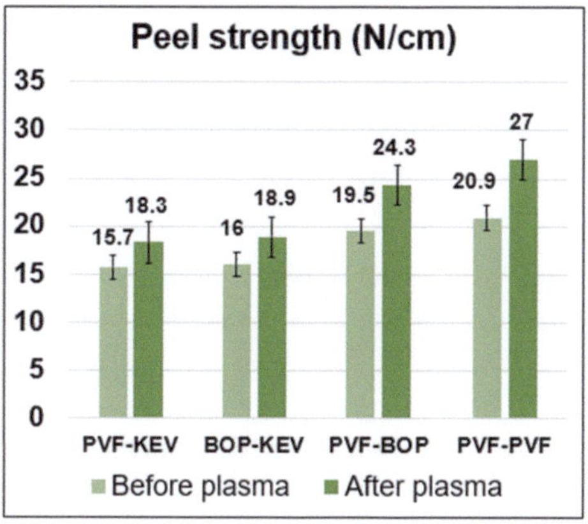

Sample code	Enhance % (after plasma)
PVF-KEV	16.6
BOP-KEV	18.1
PVF-BOP	24.6
PVF-PVF	29.2

Fig. 16 Peel strength of different double-layered laminates made from Kevlar fabric and different films, showing data before and after plasma

5.4 Effect of Plasma Treatment on Peel Strength

Figure 16 shows the change in peel strength of different double-layered laminates prepared from Kevlar fabric and two types of polymeric films with the effect of atmospheric plasma applied on the surface of the polymeric film. Before plasma treatment, the peel strength of the PVF-Kevlar joint and BOPET-Kevlar joint was 15.7 N/cm and 16 N/cm, respectively. As can be seen, plasma treatment results in significant improvement in peel strength. There was an enhancement of 16–18% in peel strength for Kevlar-based double-layered laminates.

Figure 17 shows the change in peel strength for Spectra-based double-layered laminates with the effect of atmospheric plasma. Before plasma treatment, the peel strength of the PVF-Spectra joint and BOPET-Spectra joint was 13.6 N/cm and 13.8 N/cm, respectively. As can be seen, there was a significant enhancement in peel strength (14–17%) for Spectra-based double-layered laminates. A somewhat better increment in peel strength (25–30%) was observed for peel strength between two polymeric films rather than fabrics.

5.5 Effect of Accelerated Weathering on Peel Strength

Figure 18 shows the change in peel strength of different double-layered laminates based on Kevlar fabric with different polymeric films and also shows the variation of peel strength with UV exposure. Before artificial weathering, the peel strength of PVF-Kevlar joint and BOPET-Kevlar joint was 18.3 N/cm and 18.9 N/cm, respectively.

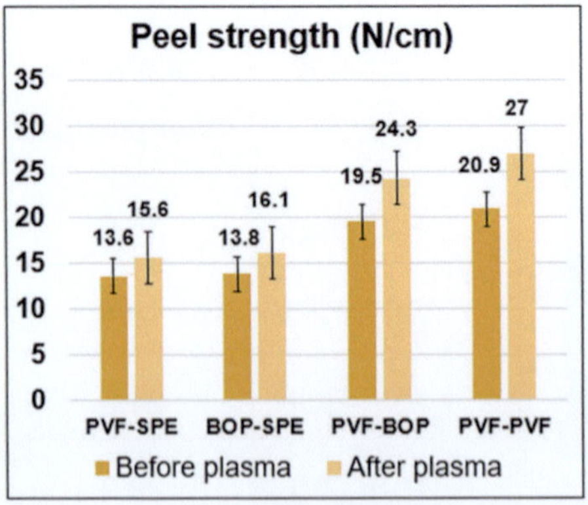

Sample code	Increase % (after plasma)
PVF-SPE	14.7
BOP-SPE	16.7
PVF-BOP	24.6
PVF-PVF	29.2

Fig. 17 Peel strength of different double-layered laminates made from Spectra fabric and different films, showing data before and after plasma

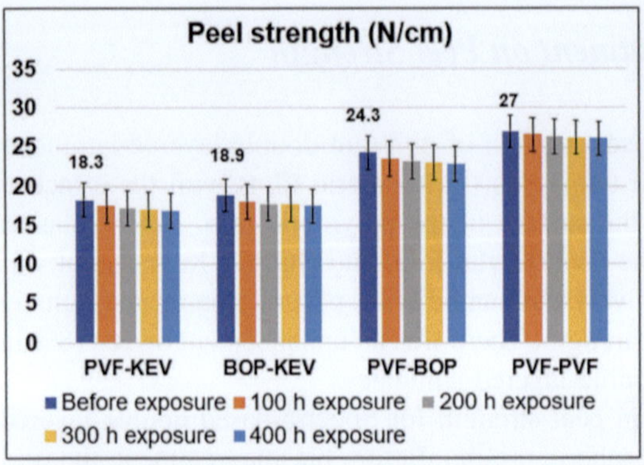

Sample code	Decrease %		
	After 100 h	After 300 h	After 400 h
PVF-KEV	4.4	6.6	7.7
BOP-KEV	4.2	6.3	7.4
PVF-BOP	3.3	5.3	6.2
PVF-PVF	1.5	3.0	3.3

Fig. 18 Peel strength of different laminates made from Kevlar fabric and different films, showing data before and after weathering

The decreased percentage of peel strength was gradually reduced from 100 to 400 h UV exposure. After 400 h exposure, decreased percentage of peel strength was 7.7, 7.4, 6.2, and 3.3% in case of PVF-Kevlar joint, BOPET-Kevlar joint, PVF-BOPET joint, and PVF-PVF joint, respectively. It can be seen that no such change was observed for the PVF-PVF joint from 300 to 400 h exposure. After 400 h UV exposure, no significant decrease in peel strength was found thereby resulting in less degradation of PU-based adhesive. Hence, the double-layered joints were stabilized after 100 h UV exposure because the maximum loss in peel strength was observed for the interval of the first 100 h exposure.

Figure 19 shows the change in peel strength of different double-layered laminates based on Spectra fabric with different polymeric films and also shows the variation of peel strength with UV exposure. After 200 h UV exposure, decreased percentage of peel strength was 7.7%, 6.8%, 4.5%, and 3.3% in case of PVF-Spectra joint,

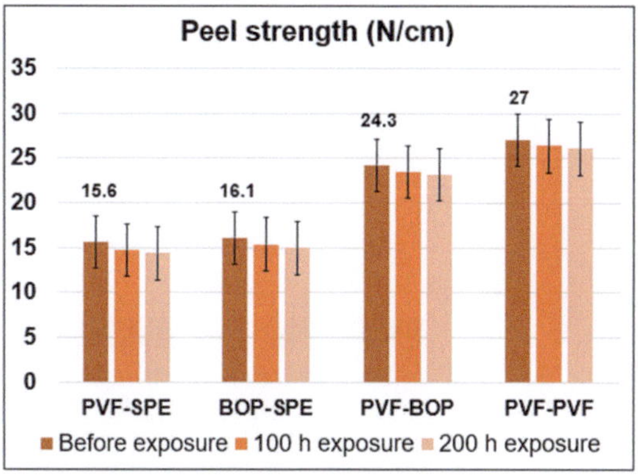

Sample code	Decrease %	
	After 100 h	After 200 h
PVF-SPE	5.1	7.7
BOP-SPE	4.3	6.8
PVF-BOP	3.2	4.5
PVF-PVF	2.1	3.3

Fig. 19 Peel strength of different (double-layered) laminates made from Spectra fabric and different films, showing data before and after weathering

BOPET-Spectra joint, PVF-BOPET joint, and PVF-PVF joint, respectively. The loss percentage in peel strength from 100 to 200 h exposure was quite less than that of exposure from unexposed condition to 100 h exposure. Hence, the double-layered joints were stabilized after 100 h UV exposure.

5.6 *Effect of Water Soaking on Peel Strength*

Figure 20 shows the variation in peel strength of Kevlar-based double-layered laminates with respect to water soaking at different temperatures. There were three different types of water soaking baths: (a) 5 °C temperature for a time period of 24 h, (b) 35 °C temperature for a duration of 24 h, and (c) 65 °C temperature for a dwell period of 4 h. Laminate specimens were immersed at least 2 inches below the water surface within the soaking bath. After completion of the soaking period, all the samples were dried in the oven at a temperature of 60 °C and then subjected to prepare for measuring the peel strength.

The water soaking test may confirm the integrity of adhesion between interface of different layers. As can be seen, the change in peel strength was very less in case of soaking at 5 °C and 35 °C temperature, whereas it was only 2–3% decrease (i.e., also insignificant loss) in case of 65 °C temperature. Hence, the interface adhesion between double-layered laminates was improved and the chances of delamination were minimum.

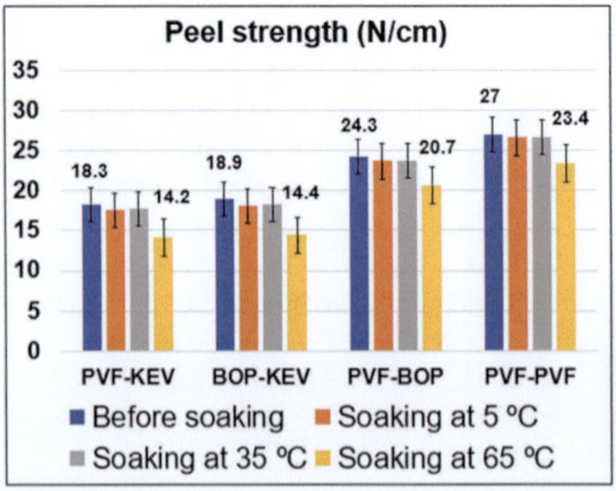

Soaking Temp (°C)	Soaking Time (h)	Drying Temp (°C)
5	24	60
35	24	60
65	4	60

Specimens are immersed at least 2 in. below the water surface followed by drying

Fig. 20 Peel strength of different (double-layered) laminates made from Kevlar fabric and different films, showing data before and after water soaking

5.7 *Effect of Accelerated Weathering on Tensile Properties*

Figure 21 shows the change in tensile strength of Kevlar fabric and Kevlar-based laminates with 100 h to 400 h UV exposure. The tensile strength of neat Kevlar fabric was 805 N/cm which was reduced to 794 N/cm after 400 h UV exposure. K2 laminate showed better tensile strength (820 N/cm) in unexposed conditions. The tensile strength of K1 and K2 was a little bit increased than that of Kevlar fabric, whereas it was a little bit decreased in the case of K3 and K4 due to more contraction on one side of the fabric thereby resulting in a spiral kind of laminates. The tensile strength of K5 and K6 was almost similar to the tensile strength of neat Kevlar fabric.

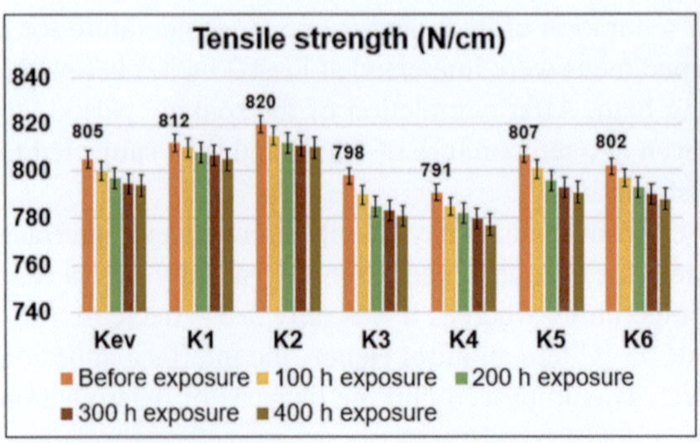

Sample code	Loss %		
	After 100 h	After 300 h	After 400 h
K1	0.3	0.6	0.9
K2	0.6	1.1	1.2
K3	1.0	1.9	2.1
K4	0.8	1.4	1.8
K5	0.7	1.7	2.0
K6	0.6	1.5	1.8

K1: PVF-Kevlar-BOPET K3: PVF-BOPET-Kevlar K5: PVF-BOPET-Kevlar-PU
K2: PVF-Kevlar-PVF K4: PVF-PVF-Kevlar K6: PVF-PVF-Kevlar-PU

Fig. 21 Tensile strength of Kevlar-based samples (Kev indicates neat Kevlar fabric, K1 indicates PVF-Kevlar-BOPET laminate, K2 indicates PVF-Kevlar-PVF laminate, K3 indicates PVF-BOPET-Kevlar laminate, K4 indicates PVF-PVF-Kevlar laminate, K5 indicates PVF-BOPET-Kevlar-PU, and K6 indicates PVF-PVF-Kevlar-PU), showing data before and after weathering

No significant decrease (only 1–3%) was observed in tensile strength for all the prepared samples, even after 400 h of UV exposure. It was confirmed that the degradation in the strength layer (i.e., fabric layer) was very less up to 400 h of artificial weathering. Hence, the tensile strength of neat Kevlar fabric was fully transferred to its laminates because of no significant loss in tensile strength among all the prepared Kevlar-based specimens.

Figure 22 shows the change in tensile strength of Spectra fabric and Spectra-based laminates with 100 h to 200 h UV exposure. The tensile strength of neat Spectra fabric was 735 N/cm which was reduced to 718 N/cm after 200 h UV exposure thereby resulting in a 2.3% loss in tensile strength. S2 laminate showed better tensile strength (763 N/cm) among Spectra-based laminates in unexposed conditions. The tensile strength of S1 and S2 was a little bit increased than that of Kevlar fabric, whereas it was a little bit decreased in the case of S3 and S4 due to more contraction on one side of the laminate.

Even after 200 h of UV exposure, no significant decrease (1–2%) was observed in tensile strength for all the prepared samples thereby resulting in less degradation in the strength layer of the laminates. As can be seen, the tensile strength of neat Spectra fabric was fully transferred to its laminates because of no significant loss in tensile strength among Spectra-based laminates. However, similar kind of trends was observed in tensile properties irrespective of laminates (Kevlar-based or Spectra-based). The laminates based on Kevlar showed greater tensile strength than that of Spectra-based laminates.

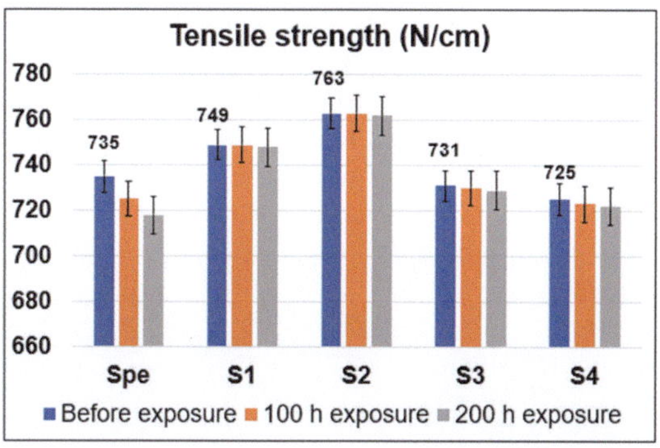

Sample code	Loss %	
	After 100 h	After 200 h
Spe	1.4	2.3
S1	0.3	0.6
S2	0.6	1.1
S3	1.0	1.9
S4	0.8	1.4

Spe: neat Spectra fabric
S1: PVF-Spectra-BOPET
S2: PVF-Spectra-PVF
S3: PVF-BOPET-Spectra
S4: PVF-PVF-Spectra

Fig. 22 Tensile strength of Spectra-based samples (Spe indicates neat Spectra fabric, S1 indicates PVF-Spectra-BOPET laminate, S2 indicates PVF-Spectra-PVF laminate, S3 indicates PVF-BOPET-Spectra laminate, and S4 indicates PVF-PVF-Spectra), showing data before and after weathering

5.8 *Effect of Accelerated Weathering on Tearing Strength*

Figure 23 shows the data of cut slit tearing strength of uncoated and adhesive-coated Kevlar fabric and different laminates based on Kevlar fabric after 400 h of UV exposure. The cut slit tearing strength of uncoated Kevlar fabric was 321 N/cm in unexposed conditions, whereas it is increased to 511 N/cm in the case of adhesive-coated Kevlar fabric.

However, the actual tearing strength of neat fabric did not get initially due to slippage of the sample during measuring the tearing strength of the uncoated fabric. When the tearing property of adhesive-coated fabric was evaluated, there was no such slippage occurred thereby the actual tearing strength of the fabric was achieved.

K1 laminate was showing better tearing strength (477 N/cm) among Kevlar-based laminates in unexposed conditions. The tearing strength of neat Kevlar fabric significantly increased (almost 60%) after applying the adhesive to the fabric because PU-based adhesive has higher flexibility and it helps in increasing the slippery effect resulting in enhancing the thread movements. The tearing strength of laminates was decreased than that of adhesive-coated fabric because the flexibility may be reduced after making laminates by compression thereby resulting in hindering the thread movements. The prepared Kevlar-based laminates provided only 1–4% tearing loss even after 400 h of UV exposure thereby resulting in less degradation of the substrate.

Figure 24 shows the data of cut slit tearing strength of uncoated and adhesive-coated Spectra fabric and different laminates based on Spectra fabric after 200 h of UV exposure. The cut slit tearing strength of uncoated Spectra fabric was 351 N/cm, whereas it is increased to 480 N/cm in case of adhesive-coated Spectra fabric in unexposed condition. It was observed that tearing strength of neat Spectra fabric significantly increased (approximately 37%) after applying the adhesive to Spectra fabric. The prepared laminates were showing excellent tearing strength in the range of 400–450 N/cm. After 100 h accelerated weathering, there were a few percentages

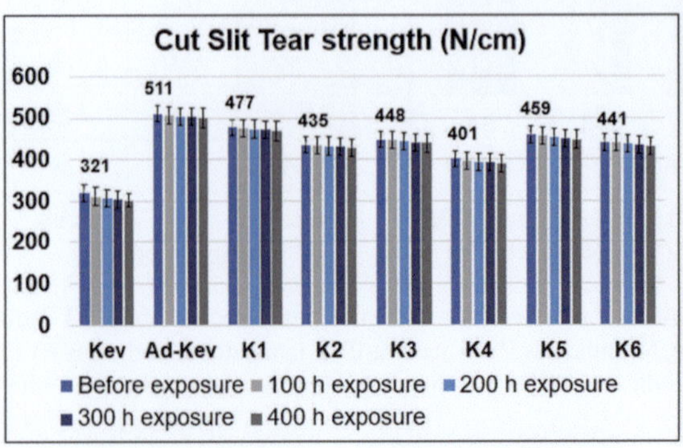

Sample code	Loss %		
	After 100 h	After 300 h	After 400 h
K1	0.4	1.3	1.9
K2	0.5	0.9	1.6
K3	0.4	1.6	2.0
K4	1.5	2.2	2.7
K5	0.6	2.0	2.6
K6	0.4	1.8	2.3

K1: PVF-Kevlar-BOPET
K2: PVF-Kevlar-PVF
K3: PVF-BOPET-Kevlar
K4: PVF-PVF-Kevlar
K5: PVF-BOPET-Kevlar-PU
K6: PVF-PVF-Kevlar-PU

Fig. 23 Cut slit tearing strength of Kevlar-based samples after 400 h weathering

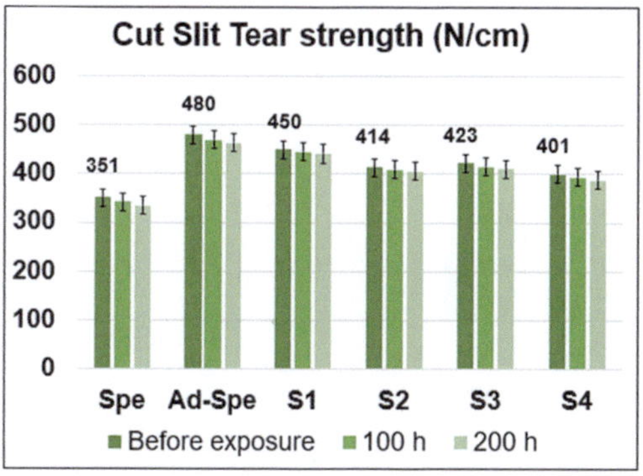

Sample code	Decrease %	
	After 100 h	After 200 h
Spe	2.56	4.84
Ad- Spe	2.08	3.33
S1	1.11	1.81
S2	1.21	1.93
S3	1.65	2.84
S4	1.74	2.99

Spe: neat Spectra fabric
Ad-Spe: Adhesive-Spectra
S1: PVF-Spectra-BOPET
S2: PVF-Spectra-PVF
S3: PVF-BOPET-Spectra
S4: PVF-PVF-Spectra

Fig. 24 Cut slit tearing strength of Spectra-based samples after 200 h weathering

of tearing loss (1–3%) observed which is not a significant loss. The same laminates provided only 2–5% tearing loss even after 200 h of UV exposure thereby resulting in less degradation of the laminated fabric.

5.9 *Effect of Accelerated Weathering on Gas Barrier Property*

Figure 25 shows the change in helium gas permeability of different laminates based on Kevlar fabric with 100 h to 400 h UV exposure. All the prepared laminates showed excellent gas barrier property (i.e., lower helium gas permeability value) irrespective of the laminate design. K5 laminate showed the best gas barrier property among Kevlar-based laminates in unexposed as well as exposed condition. The helium gas permeability of K5 laminate was 0.04 L/m^2/day at 23 °C temperature before exposure, whereas it was 0.051 L/m^2/day after 400 h UV exposure. The gas permeability was a little bit increased after 400 h exposure for all the laminates due to the degradation of the protection layer as well as the adhesive layer during exposure. As can be seen in Fig. 26, there was no such change in permeability observed from 300 to 400 h artificial weathering thereby resulting in excellent weather resistance property. The change in permeability observed for K4, K6, and K3 laminate was 19%, 20%, and 22%., respectively, thereby resulting in comparatively more degradation than that of K1 laminate during weathering. Moreover, the increased percentage in helium gas permeability for K5 laminate was maximum (~27.5%), and the values are still far lower than other laminate configurations.

Figure 26 shows the change in helium gas permeability of Spectra-based laminates with 200 h UV exposure. It can be seen that all the Spectra-based laminates showed outstanding gas barrier property (i.e., lower helium gas permeability value) as per requirements. S1 laminate showed the best gas barrier property irrespective of

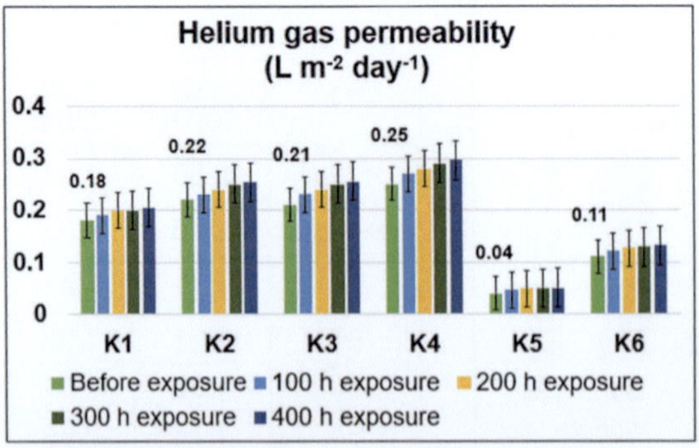

Sample code	Increase %		
	After 100 h	After 300 h	After 400 h
K1	5.6	11.1	13.9
K2	4.5	13.6	15.5
K3	9.5	19.0	21.9
K4	8.0	16.0	18.8
K5	15.0	25.0	27.5
K6	10.0	17.3	20.0

K1: PVF-Kevlar-BOPET
K2: PVF-Kevlar-PVF
K3: PVF-BOPET-Kevlar
K4: PVF-PVF-Kevlar
K5: PVF-BOPET-Kevlar-PU
K6: PVF-PVF-Kevlar-PU

Fig. 25 Helium gas permeability of different laminates based on Kevlar fabric, showing data before and after weathering

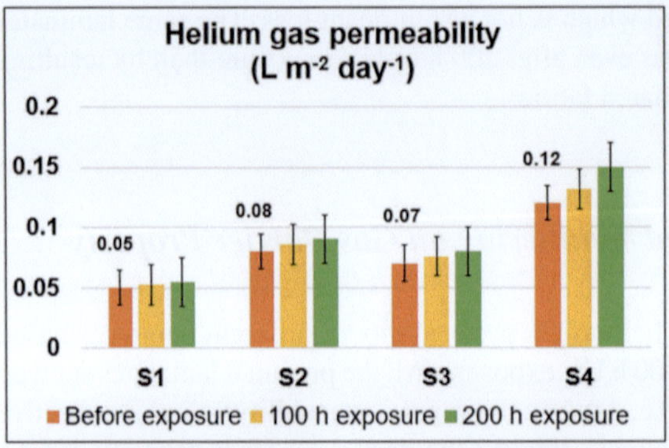

Sample code	Increase %	
	After 100 h	After 200 h
S1	6	10
S2	7.5	12.5
S3	8.6	14.3
S4	10	16.7

S1: PVF-Spectra-BOPET
S2: PVF-Spectra-PVF
S3: PVF-BOPET-Spectra
S4: PVF-PVF-Spectra

Fig. 26 Helium gas permeability of different laminates based on Spectra fabric, showing data before and after weathering

the laminate design among Spectra-based laminates in unexposed as well as exposed condition. The helium gas permeability of S1 was 0.05 L/m^2/day at 23 °C temperature before exposure, whereas it was increased to 0.055 L/m^2/day after 200 h UV exposure. It can be found that there was only a 10–17% loss observed in helium gas permeability for Spectra-based laminates.

The exposed laminates (S1, S2, S3, and S4) made from Spectra fabric showed better gas barrier property than that of unexposed laminates (K1, K2, K3, and K4) made from Kevlar fabric. Moreover, the helium gas permeability of the K5 laminate is lower than that of S1 laminate in unexposed condition thereby resulting in more potential product irrespective of the configurations of the laminates in terms of more effective gas barrier envelope.

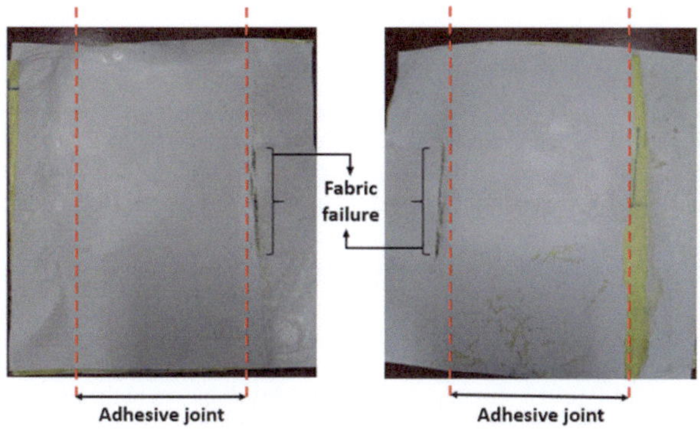

Observations	
Fabric fail	Yes
Joint fail	No
Strength	> 5000 N
Extension	3.8 %

Fig. 27 Butt joint strength of laminates (K2 laminate) after joining by adhesive sealing

6 Butt Joint Strength

Figure 27 shows the sample images (K2 laminate) for joint of two laminates and butt joint strength of laminates after joining by adhesive sealing. As can be seen, fabric failure occurred before joint failure. The fabric failure and the adhesive joint portion were observed from the sample images. The butt joint strength of the joint laminate was > 5000 N which was still far higher than breaking strength of neat Kevlar fabric thereby resulting in a good joining or good sealing made in the prepared specimen.

7 Conclusion

The current highlights show the feasibility of making multilayered laminates based on high-performance fabric with improved gas barrier and weather resistance properties. The conclusion so far of the project was given below. The contact angle and the surface energy of the films (BOPET and PVF film) were significantly decreased and increased, respectively, after plasma treatment. The crosslinker content of PU adhesive was optimized (2.5–3%). The peel strength of different double-layered joints was significantly increased (15–30%) after plasma treatment. The addition of UV additives in PU adhesive helped in increasing the peel strength as well as better retention in peel strength during UV exposure. There was no significant change in tensile properties from fabric to laminate, even after UV exposure. The tearing strength of laminates was significantly enhanced than that of fabric due to the greater flexibility of PU adhesive.

S1 was the best laminate prepared so far in terms of least helium gas permeability among Spectra-based laminates. K5 was the best laminate in terms of lower helium gas permeability and higher tearing strength among Kevlar-based laminates.

All the laminates performed better in terms of gas barrier and weather resistance properties, as compared to previously prepared PET-based laminates.

References

1. D'Oliveira FA, Melo FCLD, Devezas TC (2016) High-altitude platforms—present situation and technology trends. J Aerosp Technol Manag 8(3):249–262
2. Zhai H, Euler A (2005) Material challenges for lighter-than-air systems in high altitude applications. In: AIAA 5th ATIO and16th lighter-than-air sys tech. and balloon systems conferences, p 7488
3. Liao L, Pasternak I (2009) A review of airship structural research and development. Prog Aerosp Sci 45(4–5):83–96
4. Kang W, Suh Y, Woo K, Lee I (2006) Mechanical property characterization of film-fabric laminate for stratospheric airship envelope. Compos Struct 75(1–4):151–155
5. Stockbridge C, Ceruti A, Marzocca P (2012) Airship research and development in the areas of design, structures, dynamics and energy systems. Int J Aeronautical Space Sci 13(2):170–187
6. Chatterjee U, Butola BS, Joshi M (2016) Optimal designing of polyurethane-based nanocomposite system for aerostat envelope. J Appl Polym Sci 133(24)
7. Bagdi K, Molnár K, Sajo I, Pukánszky B (2011) Specific interactions, structure and properties in segmented polyurethane elastomers. Express Polym Lett 5(5)
8. Joshi M, Adak B, Butola BS (2018) Polyurethane nanocomposite-based gas barrier films, membranes, and coatings: a review on synthesis, characterization and potential applications. Prog Mater Sci 97:230–282
9. Chattopadhyay DK, Raju KVSN (2007) Structural engineering of polyurethane coatings for high performance applications. Prog Polym Sci 32(3):352–418
10. Kaveh P, Mortezaei M, Barikani M, Khanbabaei G (2014) Low-temperature flexible polyurethane/graphene oxide nanocomposites: effect of polyols and graphene oxide on physicomechanical properties and gas permeability. Polym-Plast Technol Eng 53(3):278–289
11. Kojio K, Furukawa M, Motokucho S, Shimada M, Sakai M (2009) Structure—mechanical property relationships for poly (carbonate urethane) elastomers with novel soft segments. Macromolecules 42(21):8322–8327
12. Špírková M, Pavličević J, Strachota A, Poreba R, Bera O, Kaprálková L, Baldrian J, Šlouf M, Lazić N, Budinski-Simendić J (2011) Novel polycarbonate-based polyurethane elastomers: Composition–property relationship. European Polym J 47(5):959–972
13. Joshi M, Bhattacharyya A (2011) Nanotechnology—a new route to high-performance functional textiles. Text Prog 43(3):155–233
14. Duncan TV (2011) Applications of nanotechnology in food packaging and food safety: barrier materials, antimicrobials and sensors. J Colloid Interface Sci 363(1):1–24
15. Benali S, Gorrasi G, Bonnaud L, Dubois P (2014) Structure/transport property relationships within nanoclay-filled polyurethane materials using polycaprolactone-based masterbatches. Compos Sci Technol 90:74–81
16. Qian Y, Lindsay CI, Macosko C, Stein A (2011) Synthesis and properties of vermiculite-reinforced polyurethane nanocomposites. ACS Appl Mater Interfaces 3(9):3709–3717
17. Park YT, Qian Y, Lindsay CI, Nijs C, Camargo RE, Stein A, Macosko CW (2013) Polyol-assisted vermiculite dispersion in polyurethane nanocomposites. ACS Appl Mater Interfaces 5(8):3054–3062
18. Awaja F, Gilbert M, Kelly G, Fox B, Pigram PJ (2009) Adhesion of polymers. Prog Polym Sci 34(9):948–968
19. Wang H (2006) Improving the adhesion of polyethylene by UV grafting. J Adhes 82(7):731–745
20. Hutchinson AR, Iglauer S (2006) Adhesion of construction sealants to polymer foam backer rod used in building construction. Int J Adhes Adhes 26(7):555–566
21. Gao M, Sun L, Guo Y, Shi J, Zhang J (2017) Modification of polyethylene terephthalate (PET) films surface with gradient roughness and homogenous surface chemistry by dielectric barrier discharge plasma. Chem Phys Lett 689:179–184

22. Pandiyaraj KN, Selvarajan V, Deshmukh RR, Gao C (2008) Adhesive properties of polypropylene (PP) and polyethylene terephthalate (PET) film surfaces treated by DC glow discharge plasma. Vacuum 83(2):332–339
23. Rezaei F, Dickey MD, Bourham M, Hauser PJ (2017) Surface modification of PET film via a large area atmospheric pressure plasma: an optical analysis of the plasma and surface characterization of the polymer film. Surf Coat Technol 309:371–381

Look Ahead Steering-Based Path Following Control for an Airship

Ramesh P. Hun and Nandan K. Sinha

1 Introduction

Airships are buoyancy-driven flying vehicles suitable for energy-efficient long endurance missions. High altitude airships are being also touted as potential alternatives for existing low-earth satellites performing surveillance and observation operations. An airship is a kind of unmanned aerial robotic platform, which processes/controls the six or more state variables in the presence of control inputs. These control inputs can be operated in many combinations depending upon the type of mission to be executed. A guidance, navigation, and control scheme is a key component in the design of an autonomous airship which makes effective use of airship control inputs. Rooz et al. [1] designed a PID-based control scheme using minimization of control efforts for long duration station keeping application of an airship. They used heading control scheme alone to keep the airship heading fixed at the set point. Development of autonomous unmanned remote monitoring airship (AURORA) is a long-term vision project led by Brazil and has made an extensive use of PID control laws for various applications of airships. De Paiva et al. [2] used a PID control approach for longitudinal velocity control and a PID-based controller for heading and altitude control for semiautonomous operation of AURORA. Carvalho et al. [3] have also designed PID-based heading controller for an unmanned robotic airship based on the H_2/H_∞ method. Other similar works related to autopilot control design for different airship models using conventional control design techniques can

R. P. Hun (✉)
Department of Electronics and Communication, Government Polytechnic for Girls, Surat, Gujarat 395001, India
e-mail: ramesh@gtu.edu.in

N. K. Sinha
Department of Aerospace Engineering, Indian Institute of Technology (IIT), Madras, Chennai, Tamil Nadu 600036, India
e-mail: nandan@ae.iitm.ac.in

D. Shukla (ed.), *Lighter Than Air Systems*, Lecture Notes in Mechanical Engineering,
https://doi.org/10.1007/978-981-19-6049-9_7

be found in [4–6]. Other useful works related to control of an airship for various objectives and using advanced control techniques are reported in [7–10]. The work reported in this article follows the procedure outlined in [9, 10] for a Gertler shape of a stratospheric airship. In Refs. [9, 10], authors considered a low altitude airship configuration and assumed all the system states, and their derivatives are available for design of control laws. However, in reality, states are available only from measurements. Systems states are estimated based on the available measurements which also help in the reduction of the on-board sensor requirements during control implementation. For implementation of LQR control law, in this work, dummy measurements of airship states based on the experimental flight results reported in [11] mixed with noise are used to generate estimates of airship states using a Kalman filter [12]. A combination of PID and LQR control schemes along with look ahead-based guidance law is used to build a complete and integrated path following guidance, navigation, and control system in MATLAB®-based simulation environment for autonomous waypoint navigation of an in-house developed stratospheric airship model [13, 14].

This paper is organized as follows: Section 2 describes the development of six degrees of freedom nonlinear mathematical model and its linearization. Section 3 presents the design of automatic flight control system. Section 4 presents the simulation results and discussion, and Sect. 5 concludes the overall work.

2 Airship Model

Airship configuration used in this paper is a stratospheric airship model developed at flight dynamics and control laboratory and is reported in [13]. A force and moment equations representing translational and rotational dynamics of airship [15] are given in Eq. (1) in the vector form.

$$M\dot{\underline{X}} = \underline{\boldsymbol{F}}_d(u, v, w, p, q, r) + \underline{\boldsymbol{A}}_a(u, v, w, p, q, r) + \underline{\boldsymbol{G}} + \underline{\boldsymbol{P}} \tag{1}$$

where M is a 6×6 mass matrix containing mass and inertia terms due to added mass or virtual mass effect. $\underline{\boldsymbol{X}}$ is the vector of state variables of airship. $\underline{\boldsymbol{F}}_d$ is the vector representing dynamic terms associated with linear and angular velocity. $\underline{\boldsymbol{A}}_a$ is 6×1 column matrix consisting of terms associated with aerodynamic forces and moments. $\underline{\boldsymbol{G}}$ is 6×1 column matrix containing the terms associated with buoyancy and gravitational forces and moments. $\underline{\boldsymbol{P}}$ is 6×1 column matrix containing the terms related to the propulsive forces and moments. The engine model considered here is kept simple and is taken from Ref. [15].

2.1 Linearization of Equations of Motion

The linearization of equations of motion requires equilibrium points around which nonlinear equations need to be linearized. The equilibrium point chosen in this linearization process is straight and level flight condition with axial trimmed velocity $u_o = 22$ m/s. The corresponding state and control vectors are

$$\underline{\boldsymbol{X}}_{eq} = [u_0, 0, 0, 0, 0, 0, 0, \theta_0]^T; \underline{\boldsymbol{U}}_{eq} = [\eta, \delta_e, 0, 0]^T \tag{2}$$

Now, the nonlinear equations of motion given by Eq. (1) are linearized around equilibrium state condition $\underline{\boldsymbol{X}}_{eq}$. This linearization process calculates the Jacobian set of nonlinear equations around the given equilibrium condition. The resulting state and control matrix is given by

$$\boldsymbol{A} = \frac{\partial \underline{F}}{\partial \underline{\boldsymbol{X}}}\bigg|_{\underline{\boldsymbol{X}}_{eq}, \underline{\boldsymbol{U}}_{eq}}, \boldsymbol{B} = \frac{\partial \underline{F}}{\partial \underline{\boldsymbol{U}}}\bigg|_{\boldsymbol{X}_{eq}, \underline{\boldsymbol{U}}_{eq}} \tag{3}$$

The corresponding linear state space model is given by

$$\delta \dot{\underline{\boldsymbol{X}}} = \boldsymbol{A}\delta\underline{\boldsymbol{X}} + \boldsymbol{B}\delta\underline{\boldsymbol{U}} \tag{4}$$

where $\boldsymbol{A}$ is a system matrix, $\boldsymbol{B}$ is a control matrix, $\delta\underline{\boldsymbol{X}}$ is an incremental state vector from equilibrium state, and $\delta\underline{\boldsymbol{U}}$ is an incremental control from equilibrium control input. The linearized state space model given by Eq. (4) can be decoupled into longitudinal and lateral directional plane dynamics for the analysis purpose.

3 Automatic Flight Control System

The automatic flight control system (AFCS) consists of guidance, navigation, and control algorithm to fulfil the autonomous mission of the airship flight. The block diagram of AFCS is shown in Fig. 1. The detailed design and description of each block are given in subsequent sections.

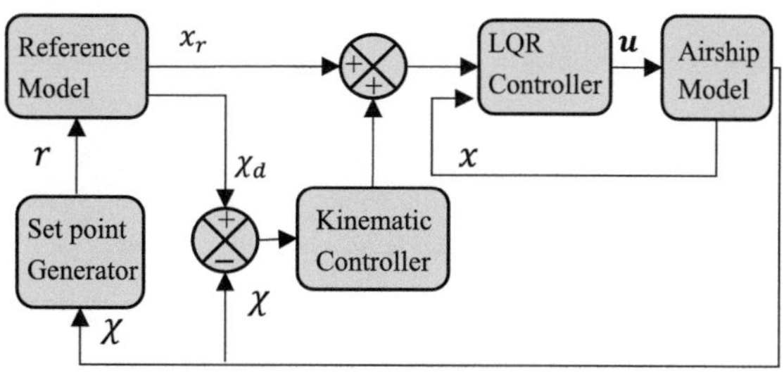

Fig. 1 AFCS block diagram

3.1 Guidance and Navigation Algorithm

The set point generator and reference model given in Fig. 1 are part of the guidance and navigation algorithm. Role of navigation algorithm is to define the current path which airship has to follow. Here, the path is created by joining the two waypoints in the flight simulation algorithm, and these waypoints are created beforehand. The navigation algorithm also switches path when required. Role of guidance algorithm in the AFCS is to ensure that airship converges to and follows the desired generated geometric path. It basically calculates the desired attitude of airship based on the current position of airship with respect to the next target waypoint. The path following problem can be solved by calculating two angles χ and υ with using look ahead steering guidance law. Graphical representation of path following scheme is shown in Fig. 2. Here, path is represented by p_i and p_{i+1} and $p(t)$ represents the current position of the airship. The objective of path following mission can be treated as

$$\lim_{t\to\infty}(p - p(t)) = 0 \tag{5}$$

Now, let us define the local frame at p_i, , known as path frame. The two consecutive elementary rotations need to be performed (using concept of Euler angles) to arrive at path frame. The first rotation is about z axis by an azimuth angle

$$\chi_p = \arctan\left(\frac{y_{i+1} - y_i}{x_{i+1} - x_i}\right) \tag{6}$$

The corresponding rotation matrix is given by

$$R(\chi_p) = \begin{bmatrix} \cos(\chi_p) & -\sin(\chi_p) & 0 \\ \sin(\chi_p) & \cos(\chi_p) & 0 \\ 0 & 0 & 1 \end{bmatrix} \tag{7}$$

The second rotation is about new y axis after first rotation by an elevation angle of

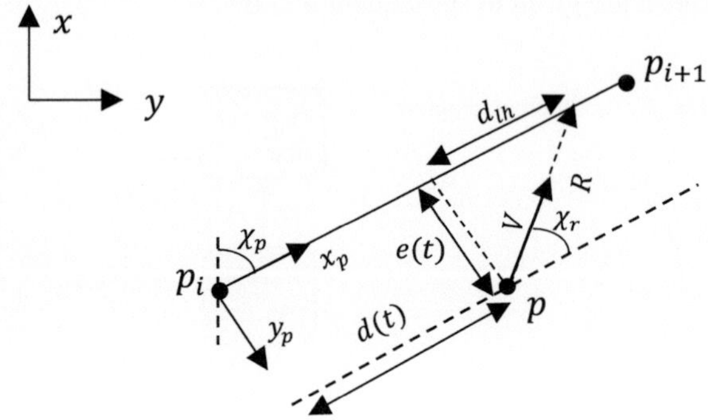

Fig. 2 NED path frame with look ahead guidance scheme

$$\upsilon_p = \arctan\left(\frac{-(z_{i+1} - z_i)}{\sqrt{(x_{i+1} - x_i)^2 + (y_{i+1} - y_i)^2}}\right) \tag{8}$$

The corresponding rotation matrix is given by

$$R(\gamma_p) = \begin{bmatrix} \cos(\upsilon_p) & 0 & \sin(\upsilon_p) \\ 0 & 1 & 0 \\ -\sin(\upsilon_p) & 0 & \cos(\upsilon_p) \end{bmatrix} \tag{9}$$

Now, the full rotation can be represented by rotation matrix

$$R_P^I = R(\chi_P)R(\upsilon_p) \tag{10}$$

Now, let us define the error vector ω^n, between airship and origin of path frame which is given by equation

$$\epsilon = R_p^I(p(t) - p_i) = R_p^I\omega^n \tag{11}$$

where ϵ is the deviation of airship path in the path frame and is given by

$$\epsilon = [d(t), e(t), h(t)]^T \tag{12}$$

where $d(t)$ is along track error, $e(t)$ is crossed track error, and $h(t)$ is vertical track error subjected to the p_i. The along track error is the distance between $p(t)$ and p_i along x axis in path frame, cross track error is the distance along the y axis, and vertical track error is the distance along z axis.

Look Ahead Steering Concept:

The look ahead steering concept shown in Fig. 2 defines the steering point towards which airship has to direct. The distance between steering point and the projection of vehicle on the path is called look ahead distance d_{lh}. The expression for look ahead distance is given by

$$d_{lh}(t) = \sqrt{R^2 - e(t)^2} \tag{13}$$

where R is a tuning variable which is running from airship position to steering point as shown in Fig. 2. This vector is known as line of sight (LOS) vector. The tuning variable R is chosen arbitrarily, and it is affecting the airship performance. The small value of R gives aggressive steering, and the large value of R gives smooth steering. The horizontal and vertical steering is defined by Eq. (14) and is also outlined in Ref. [16]. These angles are known as velocity-path relative angles, which ensure that velocity is directed towards the waypoint defined on the straight line by minimizing the cross and vertical track error.

$$\chi(e) = \chi_p + \chi_r(e);\ \upsilon(h) = \upsilon_p + \upsilon_r(h) \tag{14}$$

where

$$\chi_r(e) = \arctan\left(-\frac{e(t)}{d_{lh}}\right),\ \upsilon_r(h) = \arctan\left(\frac{h(t)}{\sqrt{e(t)^2 + d_{lh}^2}}\right)$$

3.2 Reference Model

Role of reference model used in the AFCS simulation algorithm is to generate smooth trajectory from the given series of waypoints. The set point generator generates the set of waypoints which airship has to follow during the entire flight period. When set point is changed from one set to another set, the controller faces a sudden error jump. This sudden jump in error is reduced by creating the smooth trajectory using reference model as shown in Fig. 3, and hence, controller can minimize the error more efficiently.

The reference model used in this algorithm is basically based on the mass spring damper dynamics cascaded with low-pass filter as outlined in Ref. [16]. The resulting third-order system can be represented by

$$\frac{\boldsymbol{x_r}}{\boldsymbol{r}} = \left(\frac{w_n}{s + w_n}\right)\frac{w_n^2}{s^2 + 2\zeta w_n s + w_n^2} \tag{15}$$

where x_r is a smooth reference trajectory generated from given series of set point vector r. The reference model can be tuned accordingly by tuning w_n and ζ in order to generate appropriate smooth trajectory by minimizing the error.

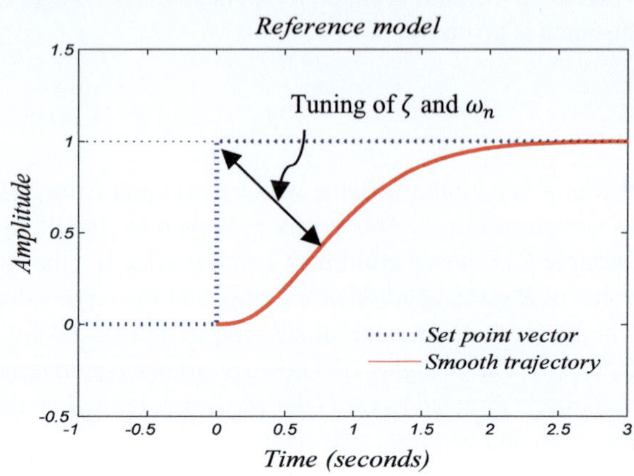

Fig. 3 Smooth trajectory using reference model

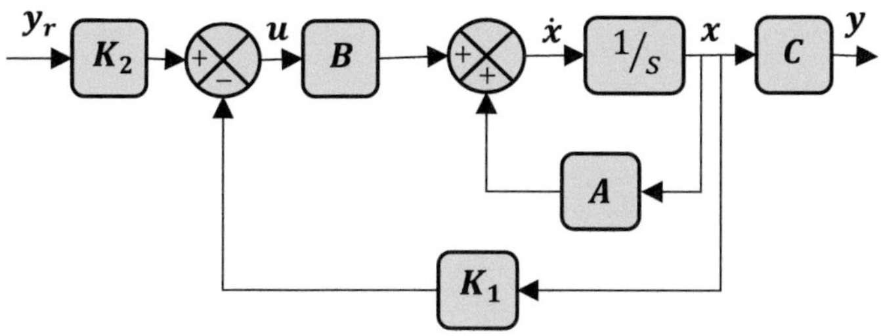

Fig. 4 LQR controller

3.3 *Control System Design*

The combination of two control strategies is designed in this section in order to execute the command signal received from look ahead-based guidance algorithm, namely (1) linear quadratic regulator (LQR) and (2) proportional, integral, and derivative (PID) controller. LQR is an optimal controller which guarantees the optimum performance of an airship by minimizing the control effort. It is basically an optimization problem which makes use of a state-feedback control law to minimize a quadratic performance index subjected to the linear state space model. However, due to state-feedback control, there will be small error which builds up with respect to the reference command. This small error in the Euler angles will turn into large angle error over the time. This large error in angles causes an inaccurate tracking in given path following mission. This problem can be overcome by designing PID control scheme because PID controller can efficiently reduce the error using feedback mechanism. Therefore, PID-based kinematic controller is designed and placed in the outer loop configuration to get rid of an error in the Euler angles.

The Design of LQR Controller:

The block diagram of LQR controller used in the inner loop configuration of automatic flight control system (AFSC) is shown in Fig. 4.

The reference signal $\boldsymbol{y_r}$ is generated from the linear reference model mentioned in AFCS block diagram and is given by

$$\dot{\boldsymbol{x}}_r = \boldsymbol{A}_r\boldsymbol{x}_r + \boldsymbol{B}_r\boldsymbol{r} \tag{16}$$

$$\boldsymbol{y}_r = \boldsymbol{C}\boldsymbol{x}_r \tag{17}$$

where $\boldsymbol{r}$ is the set point vector generated from set of waypoints in set point generator. The desired trajectory is obtained by comparing the state signal to the reference signal. The deviation of output from input is called error, and it can be minimized by tuning of w_n and ζ in the reference model. The error is given by

$$e = \boldsymbol{y} - \boldsymbol{y}_r = \boldsymbol{C}(\boldsymbol{x} - \boldsymbol{x}_r) = \boldsymbol{C}\overline{\boldsymbol{x}} \tag{18}$$

Now, LQR determines the gain vector $\boldsymbol{K}$ for the optimal control input of the form

$$\delta \boldsymbol{u} = -\boldsymbol{K}\delta\overline{\boldsymbol{x}} \tag{19}$$

which minimizes the quadratic performance index given by

$$\boldsymbol{J} = \int_0^{\infty} \left(\delta\overline{\boldsymbol{x}}^T \boldsymbol{Q}\delta\overline{\boldsymbol{x}} + \delta\boldsymbol{u}^T \boldsymbol{R}\delta\boldsymbol{u}\right)dt \tag{20}$$

where $\boldsymbol{Q}$ is a positive semi definite $n \times n$ diagonal weighting matrix, n is a number of state variable, $\boldsymbol{R}$ is a positive definite $m \times m$ diagonal weighting matrix, and m is a number of inputs. It is important to note down here that first term in the right-hand side of the performance index equation accounts for the error in the states and second term accounts for the expenditure of energy on the control signal. Therefore, the selection of $\boldsymbol{Q}$ and $\boldsymbol{R}$ is a trade-off between state error and control effort. It depends on the user and system demands whether state is important or control effort. The control command implemented here for solving the set point tracking problem is given by Eq. (21),

$$\delta\boldsymbol{u} = -\boldsymbol{K}_1\delta\overline{\boldsymbol{x}} + \boldsymbol{K}_2\boldsymbol{y}_r \tag{21}$$

Selection of $\boldsymbol{Q}$ and $\boldsymbol{R}$ matrices:

The large values of diagonal elements of $\boldsymbol{Q}$ matrix result in quick system response, while larger values of diagonal elements of $\boldsymbol{R}$ result in the more control effort. There are different methods available in the literature for selection of $\boldsymbol{Q}$ and $\boldsymbol{R}$ matrices, and author of this paper has followed Bryson's tuning rule. The Bryson's rule says that the selection of diagonal values for $\boldsymbol{Q}$ is based on the inverse of the maximum allowable value for each state, and diagonal values for $\boldsymbol{R}$ are based on the inverse of the maximum allowable value for each control input. The diagonal elements of $\boldsymbol{Q}$ and $\boldsymbol{R}$ are given by Eqs. (22) and (23), respectively.

$$q_{ii} = \frac{1}{\text{maximum allowable value of } x_i^2} \tag{22}$$

$$r_{ii} = \frac{1}{\text{maximum allowable value of } u_i^2} \tag{23}$$

After proper selection of the $\boldsymbol{Q}$ and $\boldsymbol{R}$ matrices, the LQR problem can be solved by first solving the algebraic matrix Riccati equation of the form

$$\boldsymbol{A}^T P + P\boldsymbol{A} - P\boldsymbol{B}\boldsymbol{R}^{-1}\boldsymbol{B}^T P + \boldsymbol{Q} = 0 \tag{24}$$

The solution matrix P in Eq. (24) is used to determine the optimal gain given by

$$\boldsymbol{K}_1 = \boldsymbol{R}^{-1}\boldsymbol{B}^T P;\ \boldsymbol{K}_2 = -\boldsymbol{R}^{-1}\boldsymbol{B}^T(\boldsymbol{A} + \boldsymbol{B}\boldsymbol{K}_1)^{-T}C^T\boldsymbol{Q} \tag{25}$$

The Design of Kinematic Controller:

The LQR controller is rate controller in AFSC, and hence, correction of an error in Euler angles is not included in the LQR control scheme. The proportional, integral, and derivative (PID) controller is a traditional control strategy designed based on the classical control theory. It plays a vital role in achieving the desired state of the vehicle by minimizing the error dynamics. The main purpose of designing a PID-based kinematic controller in AFCS is to minimize the error in Euler angles (pitch, yaw, and roll) by tuning of PID gains. Here, the control signal generated from the kinematic controller is a difference between desired and measured angle, and it is given by Eq. (26),

$$e = [\phi_d - \phi_m; \theta_d - \theta_m; \psi_d - \psi_m]^T \tag{26}$$

The kinematic controller is placed in the outermost loop configuration in order to prevent the error in Euler angles. In the absence of a kinematic controller, Euler angles are drifting from desired values which cause the large deviation in airship flight path from the desired path. Therefore, the kinematic controller is placed in the outer loop configuration to avoid such situation.

4 Simulation Results

The simulation results for two different flight scenarios are shown in Figs. 5 and 6. The corresponding state and control surface evolution during the entire flight is also shown in figure. The simulation is initialized in set point generator by allocating different pre-defined waypoints through which airship has to travel. The nonlinear mathematical simulation model is linearized by keeping velocity 22 m/*s* constant during the entire flight path. The corresponding equilibrium value of thrust is 6400 N. The tuning variable *R* is kept as 1200 m for the smooth turning of an airship. This tuning variable is selected based on the trial and error tuning process by observing the response characteristics. In the first scenario, total four waypoints are created in which airship has to take two turns during entire flight path. In the second scenario, a total of five waypoints are created in which airship has to come back at the same position where it was started.

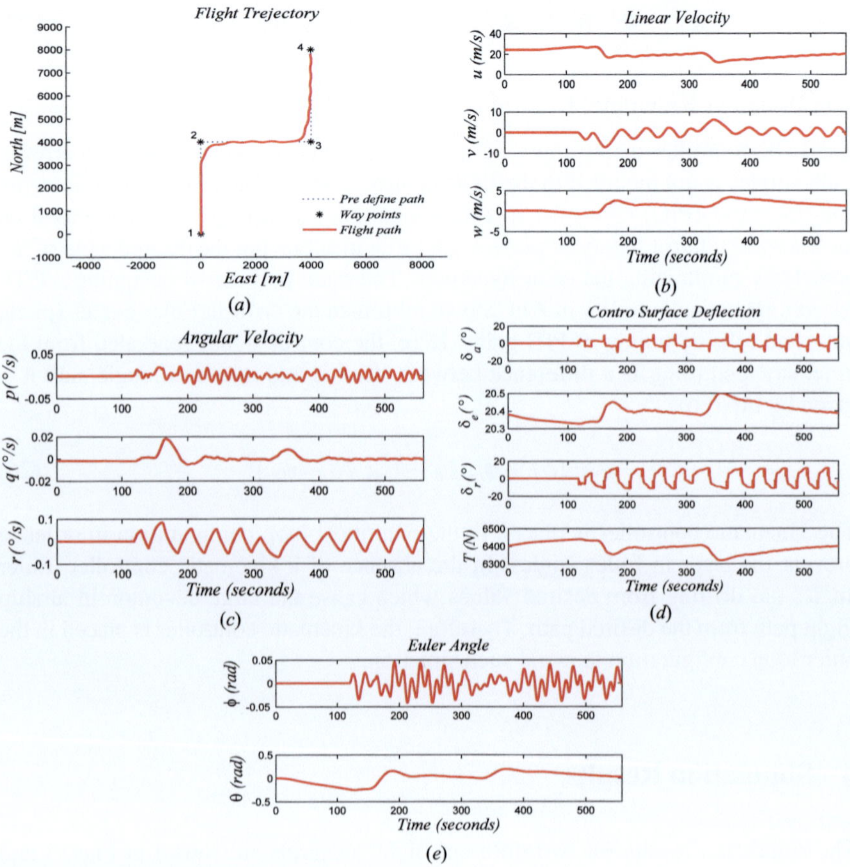

Fig. 5 Case I. Flight path with two turns **a** flight trajectory, **b** linear velocities, **c** angular velocities, **d** control surface deflection, and **e** Euler angles

5 Conclusion

This article has presented design and simulation of guidance, navigation, and control (GNC) strategy for path following problem of the stratospheric airship in simulation-based environment. The objective of GNC mission was to guide the vehicle at a constant velocity on the pre-defined planer waypoints during the entire flight path created at a particular altitude. The simulation of path following mission has included three major aspects of GNC algorithm. It includes the development of airship mathematical model, the design of guidance and navigation algorithm, and development of suitable control strategy to execute the command signal received from guidance algorithm. Look ahead-based steering guidance law is employed to direct the vehicle towards the target waypoints, while combination of LQR and PID controller is used to execute the guidance command. The performance and robustness of developed GNC

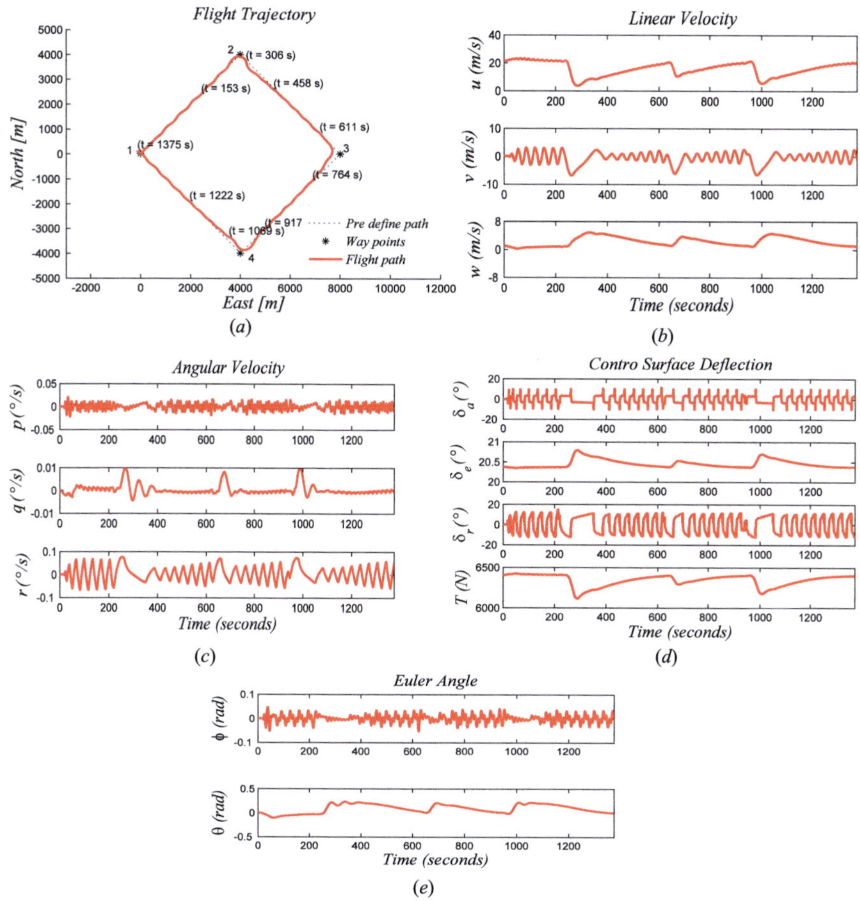

Fig. 6 Case II. Round flight path **a** flight trajectory, **b** linear velocities, **c** angular velocities, **d** control surface deflection, and **e** Euler angles

algorithm are tested in MATLAB® Simulink environment. The two different flight scenarios are presented which demonstrates the abilities and effectiveness of look ahead-based steering guidance and PID and LQR control law in the path following mission. The simulation results are presented in two different case studies which show that the excellent performance of path following mission is achieved subject to the cruise and turning flight conditions.

References

1. Rooz N, Johnson EN (2005) Design and modeling of an airship station holding controller for low cost satellite operations. In: Proceedings of the AIAA guidance, navigation, and control conference and exhibit. California, USA
2. de Paiva EC, Bueno SS, Gomes SB, Ramos JJ, Bergerman M (1999) A control system development environment for AURORA's semi-autonomous robotic airship. In: Proceedings of the IEEE international conference on robotics and automation. Michigan, USA, pp 2328–2335
3. Carvalho J, de Paiva EC, Azinheira J, Ferreira P, Ramos J, Bueno S, Bergerman M, Maeta S, Mirisola L, Faria B, Elfes A (2001) Classic and robust pid heading control of an unmanned robotic airship. In: Proceedings of the international symposium on intelligent robotic systems. Toulouse, France, pp 61–70
4. do Valle RC, Menegaldo LL, Simões AM (2014) Guidance, avionics and navigation system of an autonomous airship for demining. In: Proceedings of the 22nd international congress of mechanical engineering (COBEM), Nov 2013. Sao Paulo, Brazil, pp 2158–2169
5. Rao J, Gong Z, Luo J, Xie S (July 2005) A flight control and navigation system of a small size unmanned airship. In: Proceedings of the IEEE international conference on mechatronics and automation. Niagara Falls, Canada, pp 1491–1496
6. Mueller J (2006) Guidance, navigation and control of high-altitude airships. Princeton satellite systems, NJ, USA
7. Zheng Z, Wu Z (2013) Global path following control for underactuated stratospheric airship. Adv Space Res 52(7):1384–1395
8. Zheng Z, Huo W, Wu Z (2013) Autonomous airship path following control: theory and experiments. Control Eng Pract 21(6):769–788
9. Masar I; Stöhr E (2011) LQR-control for an autonomous airship. In: Proceedings of the 18th international conference on process control. Bratislava, Slovakia
10. Nakpiam J (2011) Control of airship motion in presence of wind, M.S. thesis, Department of mechanical and aerospace engineering, The university of Texas at Arlington, Arlington, USA
11. Liesk T (2012) Control design and validation for an unmanned, finless airship, Ph.D thesis, Department of mechanical engineering, McGill University, Canada
12. Hun R, Sinha NK, Balasubramanian P (June 2016) Design of Kalman filter for smooth state estimation of airship dynamics. National conference on large scale multi-disciplinary systems of national significance, Satish Dhawan space center, Sriharikota, India
13. Rana V (2014) Development of flight dynamic simulation model of stratospheric airship. Master degree report, Department of aerospace engineering, Indian institute of technology, Madras, India
14. Rana V, Kumar A, Sinha NK, Pal A, Sati SC (2015) Configuration analysis of stratospheric airship. Symposium on applied aerodynamics and design of aerospace vehicles (SAROD), VSSC, Trivandrum, India
15. Gomes SBV (1990) An investigation into the flight dynamics of airships with application to the YEZ-2A, Ph.D. thesis, College of aeronautics, Cranfield university, UK
16. Dybsjord KA (2013) Fault-tolerant UAV flight control system. Ph.D. thesis, Department of cybernetics, Norwegian university of science and technology, Norway

Functionalization of Polyurethane-Based Adhesives with UV Additives for LTA Applications

Shuchita Tomar, Neeraj Mandlekar, Bharti Rana, Mangala Joshi, and B. S. Butola

1 Introduction

Polyurethane (PU)-based adhesives have attracted researcher's attention as they exhibit good bonding, better outdoor stability, high flexibility, and shows very high peel strength. They are widely used in various fields such as textiles, automotive, military, food packaging industries, and lighter-than-air system (LTA) inflatables. PU adhesives contain urethane group (–C=O–) which is responsible for providing excellent adhesion property. PU polymer can be thermosets and thermoplastics which contains hard segments and soft segments.

Crystalline capable hard segments have hydrogen bonding among the urethane groups and show good interchain interaction while the soft segments show crosslinking with the hard segments [1]. The diverse properties of the polymeric material can be obtained through varying the hard and soft segment ratio. PU adhesives basically contain three types of building blocks that are diisocyanate, short-chain diol, or long-chain diol. PU adhesives show wide range of properties that generally depend on the chemical structure of isocyanates, polyols, and their functionalities. Usually, the types of isocyanates used in PU adhesives are toluene diisocyanate (TDI) and methylene diphenyl diisocyanate (MDI) [2].

The LTAs such as aerostat/airship envelope material as shown in Fig. 1 should have high tensile and tear strength, high gas-barrier property, high UV resistance property, good weather resistance property, and especially high interlaminate adhesion strength or peel strength. For developing LTA envelope material, multilayered coated or laminated fabrics are required for hull structure. Hull material should possess low weight, high strength-to-weight ratio in order to increase the payload capacity, flexible at low temperature range, high tensile and tear properties to avoid catastrophic

S. Tomar (✉) · N. Mandlekar · B. Rana · M. Joshi · B. S. Butola
Department of Textile and Fibre Engineering, Indian Institute of Technology, New Delhi 110016, India
e-mail: shuchitatomar@gmail.com

D. Shukla (ed.), *Lighter Than Air Systems*, Lecture Notes in Mechanical Engineering,
https://doi.org/10.1007/978-981-19-6049-9_8

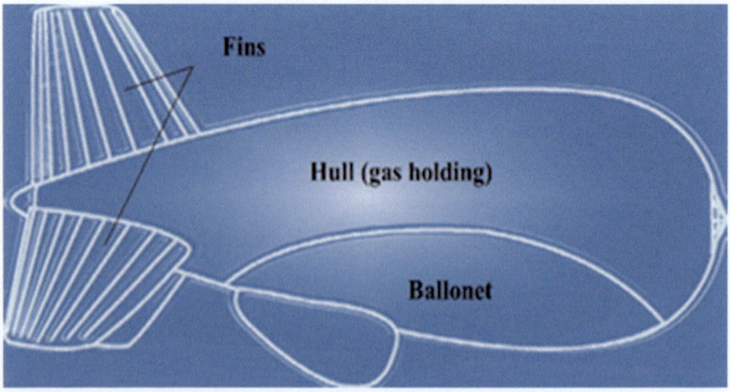

Fig. 1 Structure of LTA systems

failure ($\approx$ 2500 N and $\approx$ 313 N for aerostat material), shows good abrasion resistance and low creep behavior for better handling, bondability or sealability properties should be excellent in order to achieve good joint and leakage of gases should be prevented so that aerostat/airship life can be increased. Nowadays, generally thermoplastic polyurethane (TPU) coated nylon, Vectran, or Kevlar fabrics (plain weave or ripstop weave) are preferred because these fabrics have ability to fulfill all the abovementioned necessary requirements for designing hull structure of LTAs. TPU coated fabric reinforced with different nano fillers such as graphene, nano-clay contributes in improving the gas-barrier property of envelope material while the weather resistance property can be improved by adding UV absorber, antioxidant, hindered amine light stabilizer, carbon black, and some other inorganic nano additives such as TiO_2, ZnO, and CeO_2. Moreover, the EVOH or Mylar (DuPont, biaxially oriented polyester) film shows excellent gas-barrier property while the Tedlar (PVF) film shows good weather resistance property for multilayered laminated envelope materials. However, these polymeric films show adhesion issues with the fabric. Therefore, the interlaminar adhesion strength is one of the major challenges for aerostat/airship envelope materials [3–6].

Very less research has been carried out yet in the field of PU-based adhesives used in inflatable structure materials. Therefore, in designing such type of materials PU-based adhesive laminations were much preferred as it shows good bonding strength and high flexibility [5–10]. In this study, three types of PU-based adhesives (two-component system), i.e., Araldite, Uralane, and PLS were used and functionalized with UV additives and carbon black to enhance the UV resistant property and adhesion strength of laminates for LTA application. The Araldite adhesive usually shows fast curing and good UV stability. Uralane 5776 A/B polyurethane adhesive has a semi paste consistency for easy handling and spreading. Here, the main focus is to study the synergistic effect of functionalized adhesives on improving the interlaminate adhesion with a substrate and to optimize the GSM of the adhesive. The peel strength of the laminates was measured. Here, the substrate is the PU coated PET fabric but for better strength [7] and weather resistance property nylon, Vectran, and

Kevlar fabrics are preferred as a base material for LTA envelope. The tensile strength and UV resistance property of functionalized adhesives film were measured. Moreover, the prepared adhesive films were exposed to artificial weathering under harsh environmental conditions for 200 h. The weather resistance property, gas (helium) barrier property, and thermal properties of the adhesives were also studied.

2 Problem Definition and Formulation

Till now, the researchers are using PU-based adhesives for developing laminated structures in LTA systems in order to achieve high strength between the laminates but the interlaminate strength still need to be improved and the research is still going on to enhance the properties of LTAs. In this study, various PU-based adhesives (two-component system) were studied that are Araldite, Uralane, and PLS which exhibits high peel strength. Very few studies were done on these adhesives. Here, the adhesives were also functionalized with UV additives and carbon black in order to improve the UV resistant and weather resistance property and to optimize the adhesive gsm of laminate structures.

3 Experimental

3.1 Materials

A commercial aliphatic polyether-based thermoplastic polyurethane was purchased from Covestro, India (formerly known as Bayer Material Science Pvt. Ltd). Following are the specifications of the PU: molecular weight (Mw) = 197,200, density = 1.15gm/ cm^3, shore hardness = 85A, and refractive index = 1.503. Polyester (PET) fabric of 200 GSM (plain weave), UV additives (HALS, AO, and UV absorber) was procured from Clariant India Ltd., and carbon black nanopowder was purchased from Reinste Nano Ventures Private Ltd., dimethyl formamide (DMF), Toluene, Araldite 2028–1 adhesive (Huntsman), Uralane 5776 A/B (Huntsman), PLS adhesive, PVF film (Neoflex) gsm = 55, and BoPET film (Jindal Polyfilms) gsm = 15 (Fig. 2).

3.2 Methods

3.2.1 Process Parameters for Adhesive Lamination

Firstly, on hot-roll coating machine, the base substrate (PET fabric) was coated TPU solution then after drying the substrate tie-coat solution which consisted of TPU, UV

Fig. 2 **a** Araldite adhesive, **b** Uralane adhesive, and **c** PLS adhesive [11]

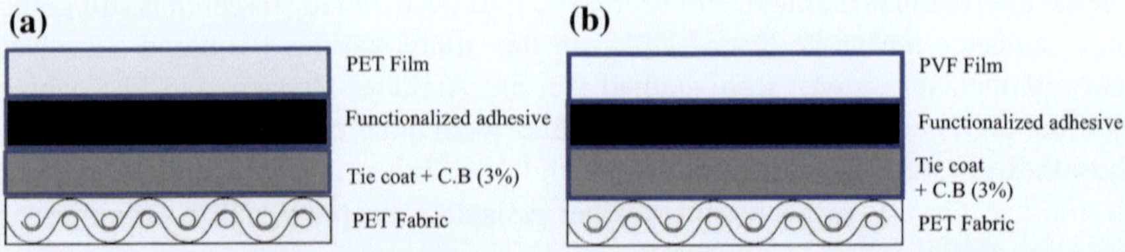

Fig. 3 Fabric-Mylar laminated structure

additives and carbon black was applied. After drying the fabric at room temperature, functionalized adhesives were applied at 0.04 mm nip roller gap to laminate the Mylar and Tedlar films using hot-roll laminator machine as shown in Fig. 3. Hence, the adhesive GSM was optimized and the peel strength was measured.

- Tie-coat layer (gap)—0.1 mm
- GSM gain—15
- Nip roller setting gap for adhesive—0.04 mm
- No temperature was given
- Speed—8.25 cm/min
- Heating was applied for drying the adhesive
- The temperature of laminated roll was set in a range of 80–85.

3.2.2 Formulation of Functionalized Adhesives

Three types of PU-based adhesives namely Araldite, Uralane, and PLS were used that are based on two-component system. In 20 ml solution of Toluene UV additives and 3% carbon black were added and sonicated for 2 h. Then, 10 gm of resin part of adhesive was mixed and stirred for 6–8 h. After that hardener was added in the solution for crosslinking and mixed thoroughly for 5 min and laminates were prepared. In case of araldite adhesive, the ratio of resin and hardener is 1:1 while in Uralane adhesive the ratio of resin to hardener is 2:1. In PLS adhesive 6% hardener was mixed (Fig. 4).

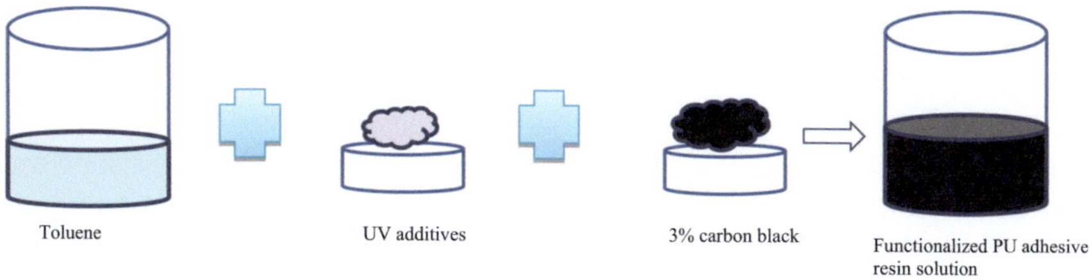

Fig. 4 Formulation of functionalized PU adhesive

3.2.3 Accelerated Artificial Weathering

The functionalized adhesive films and neat adhesive films were exposed under artificial weathering conditions in artificial weatherometer (Atlas-Xenotest 440). The test cycles were performed according to following standard: ISO 4892-2 + A1:2019. Artificial weatherometer provides weathering environment by using light sources such as filtered long arc Xenon (2000 W) so that the resistance of the film samples against weathering can be measured. The change in properties of each film with respect to the unexposed film was compared as well as analyzed after 200 h under particular exposure conditions as mentioned in Table 1.

4 Characterization and Testing

4.1 ATR-FTIR

FTIR spectroscopy (Nicolet 6700 ATR-Thermo Fisher Scientific) was used to gather FTIR spectra of neat and functionalized adhesive films before and after exposure (200 h) in accelerated artificial weathering by directly scanning the adhesive film surfaces. Number of scans taken scanning the sample surface was 64.

Table 1 Exposure conditions provided in artificial weatherometer test cycles

Conditions	Time period (min)	Irradiance (W/m^2)	Chamber temperature (°C)	R.H (%)
Dry cycle (UV)	108	0.51	35	50 ± 10
Wet cycle (UV + water)	28	0.51	35	100

4.2 Tensile Testing and Peel Strength Testing

Tensile testing of neat adhesive films and functionalized films were performed using Instron testing machine according to standard ASTM D5035. The testing was done with a gauge length of 5 cm and clamp width of 5 cm and extension rate of 300 mm/min. The peel strength was also carried out on Instron mechanical testing machine according to standard ASTM D 1876. The size of the testing sample was 12 cm × 2.5 cm.

4.3 Helium Gas Permeability

Helium gas permeability through neat PU-based adhesive films and functionalized adhesive films was evaluated by using Labthink Classic 216 gas permeation tester according to ASTM D1434 standard. The tester measures the helium gas transmission rate (GTR) through differential pressure method. The test was performed at an ambient temperature of 25 °C having 100 kPa test gas flow pressure and a constant helium gas pressure of 0.4 MPa.

4.4 Ultraviolet Protection Factor (UPF)

The ultraviolet protection factor (UPF) of neat adhesive films and functionalized adhesive films was analyzed using LABSPHERE instrument according to AS/NZS 4399:2017 testing standard method having wavelength range between 290 and 400 nm. For each sample, scanning was performed at four different places and the average values was taken to evaluate the UPF mean values.

4.5 Differential Scanning Calorimetry (DSC)

Differential scanning calorimetry (DSC) measurement was carried out on a DSC25 (TA Instruments) with in a temperature range of −60–250 °C having ramp rate of 10 °C/ min in nitrogen atmosphere and isothermal time is 1 min. The adhesive films were heated up to 250 °C, then cooled down to -60 °C and again heated up to 250 °C with same ramp rate and isothermal time.

4.6 *Thermogravimetric Analysis (TGA)*

Thermal stability of neat and functionalized adhesive films was analyzed by a TGA method (TGA55, TA Instruments). The experiment was done with a heating rate of 10 °C/min in nitrogen atmosphere in the temperature range of 50–800 °C. The flow rate of nitrogen gas was 60 ml/min.

5 Results and Discussion

5.1 *Fourier Transform Infrared Spectroscopy (FTIR) Analysis*

FTIR technique is used to identify the chemical structures of the different PU-based adhesives and functionalized PU adhesives. Carbon black contains some groups such as OH, C=O, and COOH, and these groups form hydrogen bonding with urethane linkage. It has been found that in (Fig. 5a) the presence of (1720–1730 cm^{-1}) peak shows the C=O urethane linkage in neat PLS adhesive while in functionalized PLS (1720–1730 cm^{-1}) peak almost disappear showing that the carbon black is properly dispersed in the adhesive and also the hydrogen bonding occurs between carbon black and PU-based PLS adhesive. After 200 h exposure a slight peak at (3448 cm^{-1}) shows the presence of free N–H stretching which might be due to oxidation. In (Fig. 5b), the peaks at (2850–2970 cm^{-1}) shows the symmetric and asymmetric C-H stretching of methylene groups. Peaks at (1660–1690 cm^{-1}) shows the C=O stretching, and (1505–1600 cm^{-1}) shows N–H bending. Peak band of (1104 cm^{-1}) shows C–O stretching of aliphatic ether, while in exposed Uralane film this peak becomes broad due to oxidation. In (Fig. 5c), peaks at (3300–3380 cm^{-1}) shows N–H stretching of urethane and (1660–1690 cm^{-1}) shows $C = O$ stretching. The peak intensity at (1260 cm^{-1}) shows N–H of amide linkage which slightly increases due to oxidation. Although not much change was observed under 200 h exposed Uralane and araldite adhesive film.

5.2 *Peel Strength Test*

Table 2 shows the values of adhesive gsm gain and the peel strength of laminated PET fabric. The following results are obtained after testing the peel strength of laminated fabric. In case of Uralane adhesive, better peel strength was obtained that is 7-8 N while the gain in gsm was less in case of PLS adhesive that is around 7gsm. Figure 6 shows the peel samples of laminated fabric.

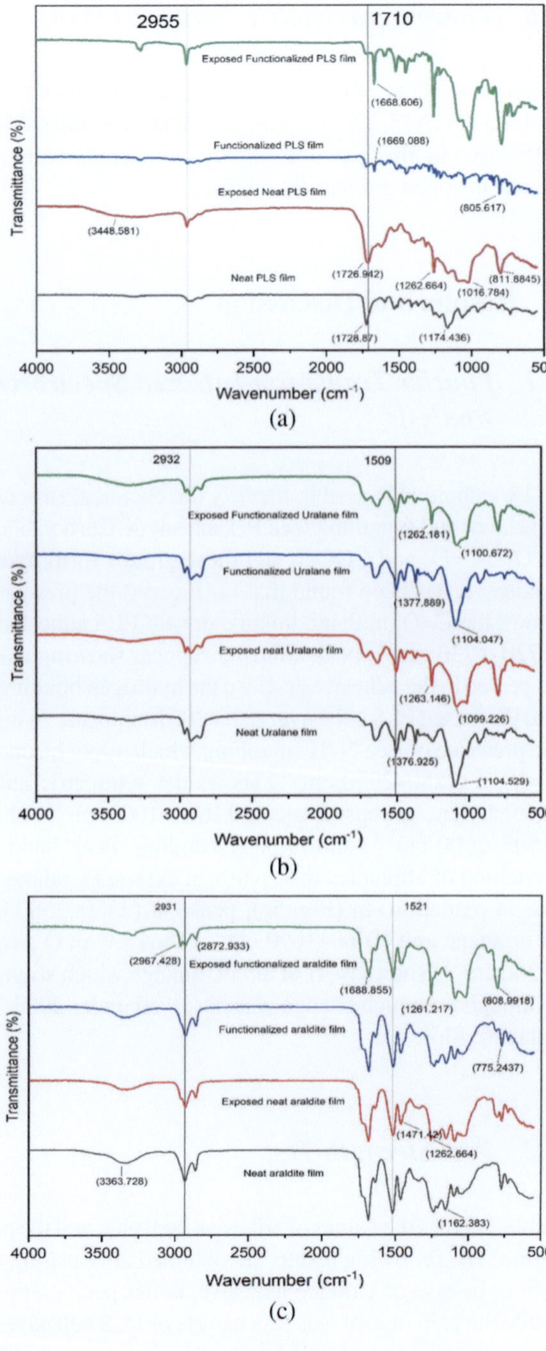

Fig. 5 FTIR spectra of unexposed and exposed neat and functionalized adhesive films **a** PLS adhesive, **b** Uralane adhesive, and **c** Araldite adhesive

Table 2 Peel strength of PU adhesives

Type of functionalized adhesive	Fabric	Layer	Final GSM	Nip roller gap (mm)	Adhesive GSM	Peel strength (N)	Breaking film
Araldite (Huntsman) Two-component based (1:1)	PET fabric (200 GSM) + Tie coat	Tedlar-fabric	312	0.04	42	6–7	Break
		Mylar-fabric	272				
Uralane (5776 A & B Huntsman) Two-component based (2:1)	PET fabric (200 GSM) + Tie coat	Tedlar-fabric	303	0.04	33	7–8	Break
		Mylar-fabric	263				
PLS 500 with hardener (Pidilite)	PET fabric (200 GSM) + Tie coat	Tedlar-fabric	277	0.04	7	5–6	Break
		Mylar-fabric	237				

(a)

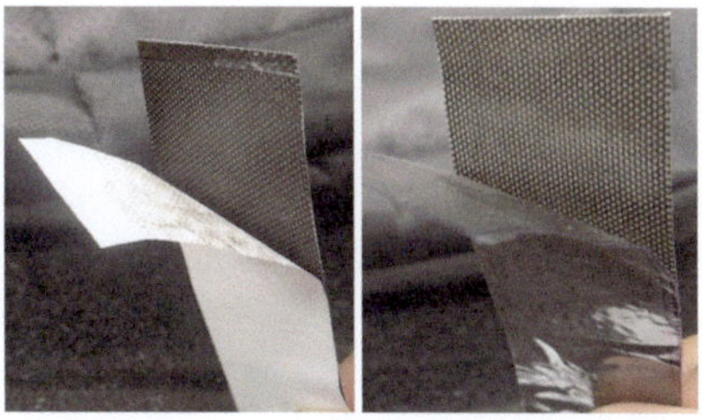

(b)

Fig. 6 **a** Laminated and coated PET fabric and **b** peel samples of laminates

5.2.1 Tensile Properties of Adhesive Films

Tensile properties of functionalized adhesive films are improved as compared to neat films as shown in (Fig. 7). Carbon black (3%) dispersed properly in the adhesives and good crosslinking as well as hydrogen bonding between the CB and urethane linkage as shown in FTIR significantly enhanced the tensile strength of functionalized adhesive films. Although when these films are exposed to artificial weatherometer under harsh environmental conditions for about 200 h the tensile strength of PLS and Uralane functionalized adhesive decreases while in araldite it increases. Similarly, Uralane and araldite neat adhesive films show increment in tensile strength

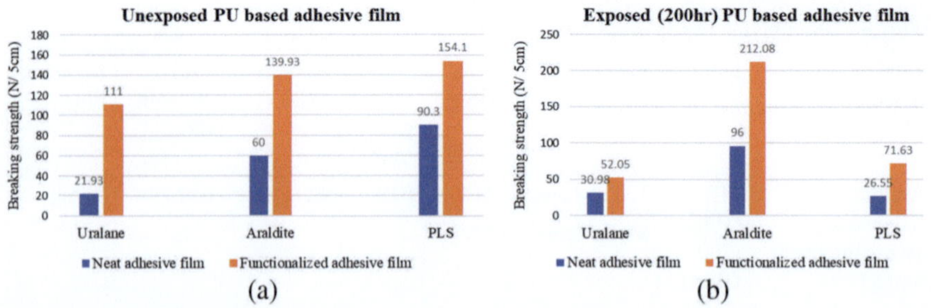

Fig. 7 **a** Tensile strength of unexposed neat and functionalized adhesive films and **b** tensile strength of exposed (200 h) neat and functionalized adhesive film

as compared to PLS adhesive. The improvement in strength is might be because of crosslinking between the PU-based adhesives. It was observed that araldite adhesive does not degrade that much within 200 h exposure.

5.3 Helium Permeability Testing

Figure 8a shows the result of helium gas permeability unexposed adhesive films. It was observed that helium permeability of unexposed functionalized adhesive film was slightly changed compared to neat adhesive film. However, when the film was exposed up to 200 h, the helium permeability was significantly improved. This behavior was attributed due to the crosslinking of intermolecular chains of PU that leads to crosslinked structure of formation but after certain time period of exposure PU-based adhesive can be degraded because of the breaking of intermolecular bond between the crosslinking chains (Table 3).

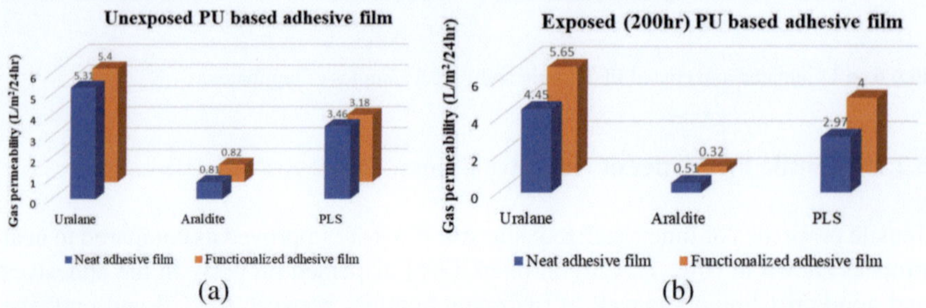

Fig. 8 **a** Helium gas permeability of unexposed neat and functionalized adhesive film; **b** helium gas permeability of exposed (200 h) neat and functionalized adhesive film

Table 3 Helium gas permeability of adhesive films

Adhesive films	Unexposed		Exposed (200 h)	
	Gas permeability ($L/m^2/24$ h)	Coefficient_P (Pc) ($cm^3.cm/cm^2$.s.cm Hg)	Gas permeability ($L/m^2/24$ h)	Coefficient_P (Pc) ($cm^3.cm/cm^2$.s.cm Hg
Neat Uralane film	5.31	1.21×10^{-9}	4.45	1.42 × 10–9
Functionalized Uralane film	5.4	1.24×10^{-9}	5.65	1.81×10^{-9}
Neat PLS film	3.46	7.92×10^{-10}	2.97	3.22×10^{-10}
Functionalized PLS film	3.18	7.28×10^{-10}	4.0	3.57×10^{-10}
Neat araldite film	0.81	4.95×10^{-10}	0.51	3.04×10^{-10}
Functionalized araldite film	0.82	4.9×10^{-10}	0.32	3.14×10^{-10}

5.4 UV Protection Factor (UPF) Analysis

Ultraviolet protection factor (UPF) is used to measure the UV resistance property of material. It shows the amount of UV-B and UV-A radiations can be transmitted or blocked by any material. This test was performed using LABSPHERE instrument according to AS/NZS 4399:2017 testing standard. The wavelength range was 290-400 nm. UV rays blocked in UV-A and UV-B region was evaluated along with UPF mean value. UPF rating is calculated by following equation [12].

$$\mathrm{UPF} = \frac{\sum_{\lambda=280}^{400} E\lambda \times S \times \Delta\lambda}{\sum_{\lambda=280}^{400} E\lambda \times S \times T\lambda \times \Delta\lambda}$$

where

$E\lambda$—solar UVR spectral irradiance ($W \cdot m^{-2}\ nm^{-1}$).

$S\lambda$—solar spectral irradiance.

$\Delta\lambda$—wavelength interval measurement (nm).

$T\lambda$—average spectral transmittance.

Figure 9 shows the UPF values of unexposed and exposed adhesive films. UPF mean value of neat Uralane adhesive was higher as compared to araldite and PLS, while the UPF value of functionalized adhesive film was much higher as compared to neat adhesive film which shows that the interaction of carbon black with PU adhesive results in excellent UV resistant property (UPF–2000). Although neat adhesive film when exposed to 200 h in artificial weathering shows improvement in UV resistant property due to formation of cross-links between the PU chains.

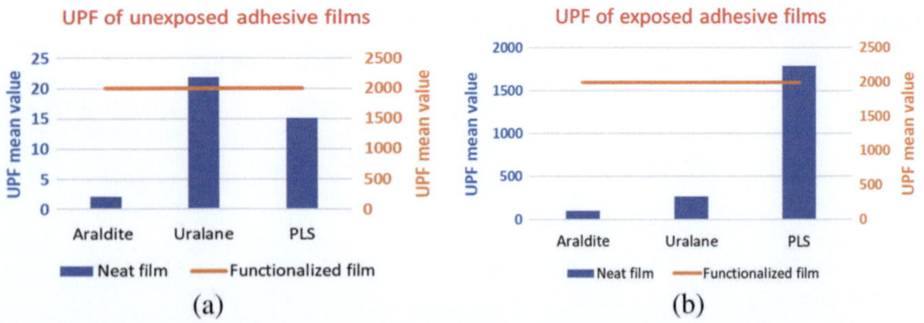

Fig. 9 UPF mean values of **a** unexposed adhesive film and **b** exposed adhesive film

5.5 *Thermogravimetric Analysis (TGA)*

Thermal stability of PU-based adhesive was analyzed through TGA technique. Figure 10a shows the TGA curve of neat and functionalized adhesive films. The 2% weight loss temperature ($T_{2\%}$) in Uralane neat adhesive was observed at 234.18 °C (onset temp—329.18 °C), while in functionalized Uralane $T_{2\%}$ was 227.76 °C (onset temp—333.76 °C); whereas in neat araldite adhesive was 109.56 °C (onset temp—300 °C) and 223.09 °C (onset temp—298.82 °C) was observed in functionalized araldite adhesive. In PLS, neat adhesive ($T_{2\%}$) was 266.88 °C (onset temp—350.86 °C) and 188.78 °C (onset temp—339.62 °C). The following trend was observed due to the addition of carbon black and UV additives which are less thermally stable as compared to neat PU-based adhesives.

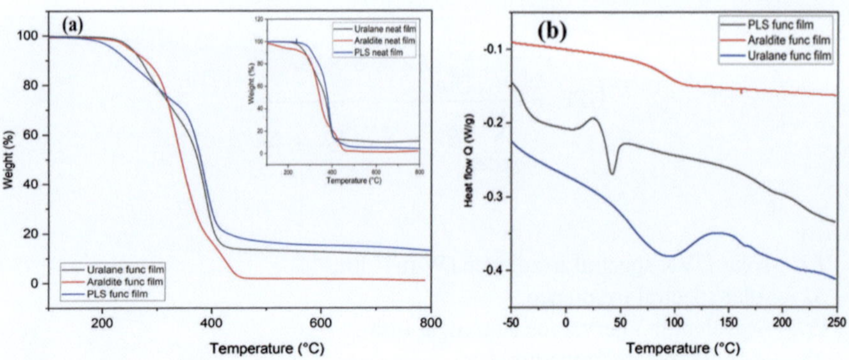

Fig. 10 **a** TGA curve for neat and functionalized adhesive film; **b** DSC thermograms for functionalized adhesive film

5.6 *Differential Scanning Calorimetry (DSC) Analysis*

Figure 10b represents the DSC thermogram of functionalized PU-based adhesive. In PLS functionalized adhesive, an endotherm peak was observed at 40 °C which relates to the glass transition temperature of PLS (T_g); while in Uralane and araldite functionalized adhesive an endothermic peak was shifted to higher temperature between 85–120 °C which shows the T_g of hard segments of adhesive.

6 Conclusion

In LTA systems, adhesion strength between the laminates must be higher. In this study, three types of PU-based adhesive (Uralane, Araldite, and PLS) were used and functionalized by using carbon black (3%) and UV additives. The peel strength of laminated fabric in functionalized Uralane adhesive was very high (7–8 N). The gsm gain was less in case of functionalized PLS and also adhesion strength was better. The hydrogen bonding between the carbon black and PU-based adhesives significantly improves the tensile properties of functionalized adhesive films especially PLS shows good breaking strength. Functionalized adhesives films show excellent UPF values (2000) which led to improve the weather resistance property also. However, thermal stability of functionalized adhesives decreases slightly because carbon black and other UV additives start decompose at lower temperature as compared to PU-based adhesives. Moreover, when prepared films were exposed to artificial weathering up to 200 h the functionalized adhesive shows excellent weather resistance behavior. Thus, functionalized PLS and Uralane adhesives can fulfill the requirement for LTA laminated structures as compared to araldite adhesive. Therefore, these adhesives can be preferred in the future for the preparation of multilayered coated and laminate structures for LTAs.

Acknowledgements The authors are very grateful to Joint Advanced Technology Centre (JATC), DRDO India and IIT Delhi to provide research facilities for the completion of this research work.

References

1. Dodiuk H, Belinski I, Dotan A, Kenig S (2006) Polyurethane adhesives containing functionalized nanoclays. J Adhes Sci Technol 20(12):1345–1355. https://doi.org/10.1163/156856106778456573
2. Sheikhy H, Shahidzadeh M, Ramezanzadeh B, Noroozi F (2013) Studying the effects of chain extenders chemical structures on the adhesion and mechanical properties of a polyurethane adhesive. J Ind Eng Chem 19(6):1949–1955. https://doi.org/10.1016/j.jiec.2013.03.008
3. Armağan OG, Kayaoğlu BK, Karakaş HC, Güner FS (2013) Improving the adhesion strength of polypropylene nonwoven laminated fabrics Using low-pressure plasma. Fibres Text East

4. Pizzi A (2006) Recent developments in eco-efficient bio-based adhesives for wood bonding: Opportunities and issues. J Adhes Sci Technol 20(8):829–846. https://doi.org/10.1163/156856106777638635
5. Dasaradhan B, Ranjan Das B, Kumar Sinh M, Kumar K, Kishore B, Eswara Pra N (2017) A brief review of technology and materials for aerostat application. Asian J Textile 8(1), 1–12, Dec. https://doi.org/10.3923/ajt.2018.1.12
6. Kumar A, Sati SC, Ghosh AK (2016) Design, testing, and realisation of a medium size aerostat envelope. Def Sci J 66(2):93–99. https://doi.org/10.14429/dsj.66.9291
7. Mishra D, Kumar Sinha V (2010) Eco-economical polyurethane wood adhesives from cellulosic waste: synthesis, characterization and adhesion study. Int J Adhesion Adhes 30(1), 47–54, Jan. https://doi.org/10.1016/j.ijadhadh.2009.08.003
8. Adak B, Joshi M (2018) Coated or laminated textiles for aerostat and stratospheric airship
9. Mahesh V, Joladarashi S, Kulkarni SM (2019) An experimental study on adhesion, flexibility, interlaminar shear strength, and damage mechanism of jute/rubber-based flexible 'green' composite. J Thermoplast Compos Mater. https://doi.org/10.1177/0892705719882074
10. Raza W, Singh G, Kumar SB, Thakare VB, Challenges in design & development of envelope materials for inflatable systems [Online]. Available: www.tjprc.org
11. Pidilite Polyurethane Adhesive—FEVICOL PLS 500, Tin Can, Rs 215/litre | ID: 4831464197. https://www.indiamart.com/proddetail/polyurethane-adhesive-fevicol-pls-500-4831464197.html. Accessed 07 Sept 2022
12. Louris E et al (2018) Evaluating the ultraviolet protection factor (UPF) of various knit fabric structures. IOP Conference Series: Materials Science and Engineering 459(1). https://doi.org/10.1088/1757-899X/459/1/012051

Degradation Study of Aromatic and Aliphatic TPU Films in Accelerated Weathering: Impact on the Gas Barrier and Mechanical Properties

Neeraj Mandlekar, Rishabh Tiwari, Sampath Parasuram, and Mangala Joshi

1 Introduction

Thermoplastic polyurethane (TPU) films and coatings are widely used for outdoor structures, inflatable systems and lighter than air (LTA) vehicles/systems due to a variety of excellent properties. TPU is also used as aerostat and airship design material [1]. However, like other polymeric materials, TPU is also susceptible to degradation when exposed to aggressive environments which significantly deteriorate the properties of TPU. Prolonged exposure under natural or artificial weathering causes irreversible degradation in chemical structure, mechanical properties and physical properties of PU films due to the presence of harmful UV radiation in the solar spectrum. As a consequence, the helium gas barrier and tensile properties of PU films are significantly affected and the overall service life of the LTA envelope is reduced. However, the weather resistance properties of PU films strongly depend on chemical structure and molecular chain orientation. It has been reported that aliphatic grade PU provides better weather resistance than aromatic grade PU [2].

In the last few years, the influence of the weathering process on the prediction of life of the envelope material has attracted increasing attention [3]. In this context, several studies have been proposed doing natural and artificial weathering tests [4–6]. However, natural weathering is most of the time is not repeatable due to dependency on several natural factors. Therefore, the artificial weathering test is more reliable to simulate natural weathering in controlled weatherometer apparatus [7, 8]. It has been reported that xenon arc-based light sources are widely used to simulate UV radiation from the Sun, and such artificial weathering has been demonstrated in several studies on the ageing characteristic of polymers and polymeric blends/composite s [9]. There are a few studies which have been carried out with polyurethane films and

N. Mandlekar (✉) · R. Tiwari · S. Parasuram · M. Joshi
Department of Textile and Fibre Engineering, Indian Institute of Technology Delhi, Hauz Khas, New Delhi 110016, India
e-mail: neeraj.mandlekar@textile.iitd.ac.in

D. Shukla (ed.), *Lighter Than Air Systems*, Lecture Notes in Mechanical Engineering,
https://doi.org/10.1007/978-981-19-6049-9_9

coatings in xenon arc-based accelerated artificial weathering conditions using laboratory weatherometer [10, 11]. However, weathering environment has not been fully described using xenon lamp because it produces a higher relative intensity of shorter wavelength which simulate the extremely harsh environment more accurately. Thus, to simulate outdoor natural weathering in artificial weatherometer apparatus more environmental factors have to be incorporated.

Hence, in this work, accelerated artificial weathering process is particularly designed for testing of coated and laminated flexible textile material which is used in developing envelope of LTA systems. Such weathering condition especially gives the UV radiation in the range of 340 nm using glass filters, other weathering parameters like irradiance, humidity, temperature and wind speed are set to simulate natural weathering with a higher UV exposure rate. To study weather resistance properties aliphatic and aromatic polyether-based different types of TPU have been used to study the influence of accelerated artificial weathering on tensile properties, helium gas barrier and change in surface morphology.

2 Material and Method

2.1 Materials

Aromatic and aliphatic polyether-based thermoplastic polyurethane (TPU) polymer resins were procured from Lubrizole Advanced Materials, Europe, under the trade name of ESTANE®. Pellets were dried under vacuum for 3 h at 70 °C before transforming them into a continuous film.

2.2 TPU Extrusion Cast Film Preparation

Polyether-based aliphatic and aromatic TPU films used in this study were produced in the lab using a single screw extruder connected with cast film unit (LabTech Engineering, Model LCR 300 HD, Thailand). Single screw L/D 30 was used at predefined temperature profile 160, 170, 170, 180, 180 °C. Roller speed was maintained between 0.8 and 1 m/min to obtain a uniform thickness of continuous 100 μm film with 30 cm width. Figure 1 shows the single screw extruder and extrusion film casting assembly used for developing TPU films.

Fig. 1 Different stages of aliphatic and aromatic TPU film preparation by using single screw extruder connected with film casting unit

2.3 *Artificial Accelerated Weathering Test*

Artificial weathering tests were performed to evaluate the stability of the TPU films studied. ATLAS Xenotest 440 weatherometer was used, which simulates the severe outdoor environmental condition (see Fig. 2) The 2000 W xenon arc lamp was used to produce solar radiation, and a glass filter was used to exclusively allow 340 nm UV radiation. Weathering tests were carried out according to ISO 4892-2 standard, which is used to study artificial weathering tests for coated flexible textile materials. Both the aromatic and aliphatic films were exposed in the combined effect of irradiance, temperature, humidity and wind at different time intervals of a dry and a wet cycle. Irradiance was 0.51 ± 0.01 W/m^2, temperature 38 °C $\pm$ 1, fan speed 2000 rpm and relative humidity was 50% during dry cycle and 100% in wet cycle. A dry cycle was set for 102 min and a wet cycle for 18 min. These dry and wet cycles were repeated up to 500 h. Testing samples were taken out every 100 h to investigate the mechanical properties, helium gas barrier, surface morphology and chemical structure change.

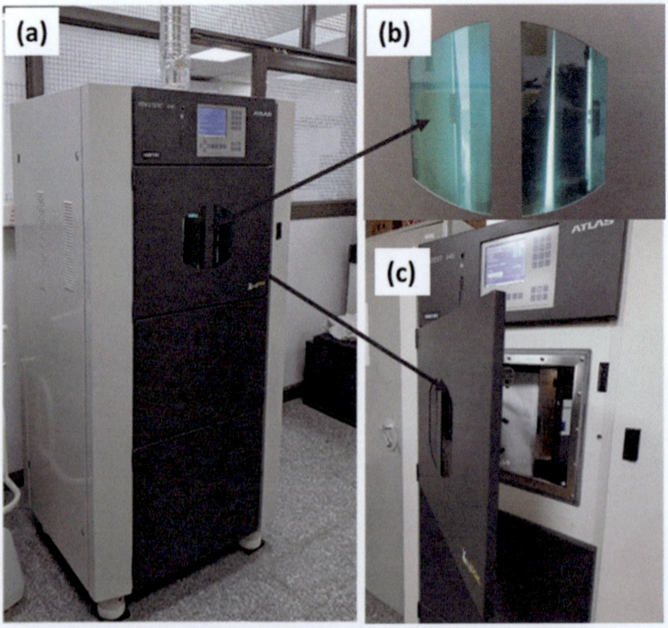

Fig. 2 Accelerated artificial weathering test: **a** laboratory weatherometer, **b** exposure chamber and **c** chamber with sample panel in spherical fashion

2.4 Characterization Techniques

Following characterization techniques were used to evaluate the weather resistance properties of the exposed TPU films.

Tensile Testing. Mechanical strength of exposed films was measured in INSTRON tensile testing machine (Model no 5960), USA. Load cell with 5 kN capacity was used, and tests were carried out in tensile mode with specimen size 10 cm × 2.5 cm keeping extension rate 500 mm/min according to ASTM standard D882. Results were extracted from Bluebill software, and breaking strength, modulus and % elongation were reported.

Helium gas permeability. Low helium gas permeability is a critical requirement of any LTA system. For artificial weathered films, helium gas permeability was measured according to ASTM D1434 using Labthink, Classic 216 gas permeability tester. Tests were carried out through the differential pressure method in proportional mode with helium gas pressure 100 kPa and test area was 38.48 cm^2. The helium permeability test was repeated three times for each sample.

ATR-FTIR analysis. Change in chemical structure was assessed by FTIR analysis in ATR mode using Nicolet iS50 Thermoscientific with ZnSe crystal. Spectra were recorded at room temperature in the range from 4000 to 500 cm^{-1} using 64 scans with 4 cm^{-1} optical resolution.

SEM analysis. Surface morphology of unexposed and exposed TPU films was studied in ZEISS Ev18 apparatus equipped with an X-ray probe (INCAEnergy Oxford, Cu Ka X-ray source, k ¼ 1.540 562 Å), under the voltage of 20 kV. The sample was cut in 2 mm^2 shapes and placed on carbon tape and sputter-coated with the gold layer.

AFM analysis. The surface roughness of exposed TPU films was assessed by AFM analysis. When the exposed films were removed from a weatherometer chamber after different exposure time intervals, a 5 mm^2 square specimen was cut for AFM studies. The AFM used in this work was a Nanoscope IIIa (Digital Instruments, California). A "J" scanner (125 μm scan region) was employed to collect the topography of 20 μm^2 region. The collected 2D AFM images were processed by Gwyddion software to obtain 3D AFM images.

UV protection factor measurement. The ultraviolet protection factor (UPF) of TPU films was measured for unexposed and exposed TPU film samples. Testing was performed according to standard AATCC 183:2000, by a UV transmittance analyser (Labsphere 2000F) over a wavelength range of 290–450 nm. For each sample, scanning was done at ten different places, and an average was taken to calculate the % UV transmittance and UPF rating.

3 Results and Discussion

3.1 Mechanical Properties

Tensile properties of TPU films evaluated before and after the exposure shows that the aromatic and aliphatic TPU films behave differently during tensile testing (see Fig. 3). Tensile testing results are collected in Table 1. In aliphatic TPU films as the exposure time increases breaking strength and % elongation abruptly increase in 100 h due to the formation of stronger bonds which allows to better stress transfer resulting in higher breaking strength. Further, increasing the exposure time gradually reduces the breaking strength and % elongation, at the same time modulus increases up to 500 h but after that, film material degraded and start behaving like soft material and all tensile parameters decreased. This is ascribed to the scission of longer polymer chains into shorter chains due to the degradation under artificial weathering (UV exposure). Besides, aromatic TPU films also show similar behaviour in 100 h and increment in breaking strength, % elongation noticed thereafter increasing the exposure time lower the breaking strength and % elongation, at the same time, modulus increases with longer exposure time. This behaviour is attributed to the crosslinking in the polymer chains caused by UV radiation. After 300 h of exposure, aromatic TPU film becomes brittle material because the modulus value significantly increases but other

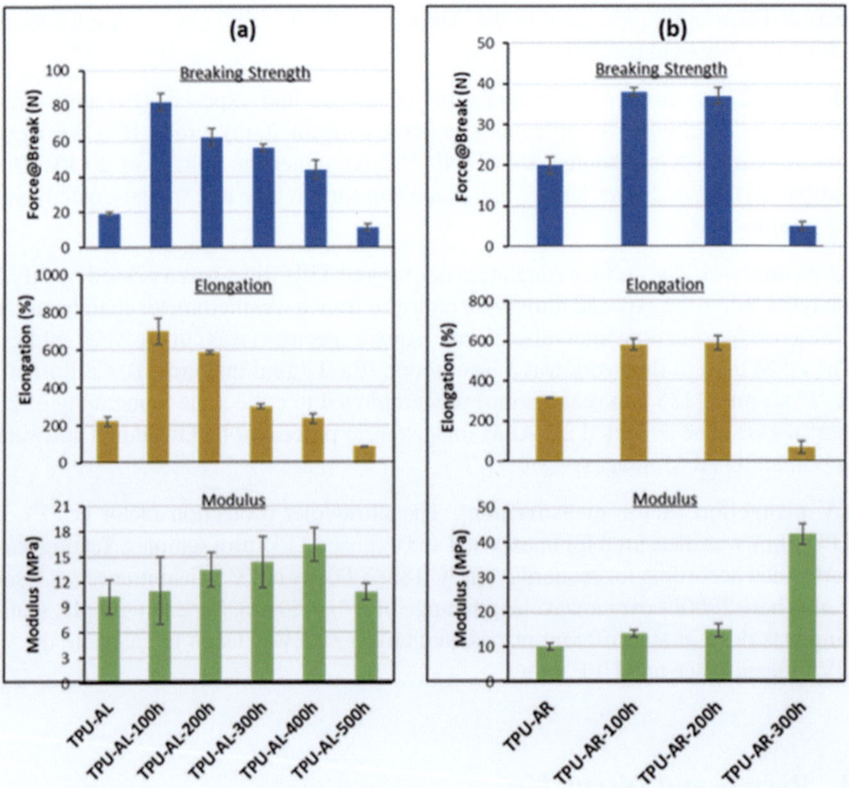

Fig. 3 Tensile properties of aliphatic and aromatic TPU films after exposure; (a) TPU-AL and (b) TPU-AR films

tensile parameters decreased which is attributed to the crosslinking behaviour in aromatic TPU film. This crosslinking behaviour is also confirmed by FTIR analysis.

3.2 Helium Gas Permeability

Helium gas barrier is one of the most important properties of LTA envelope material, and TPU films are used in designing weather resistance and gas barrier layers material. Helium gas permeability study of aliphatic and aromatic TPU exposed film exhibit opposite behaviour to each other. Helium permeability of artificial weathered TPU films is shown in Fig. 4a–b and data is collected in Table 2. In order to get extrapolated values for a longer duration of exposure, the polynomial curve fitting equation was used, and data was extrapolated up to 900 h exposure time. Helium barrier results of TPU-AL film show that permeability was slightly changed up to 500 h exposure. This behaviour shows that aliphatic film has weather resistance

Table 1 Tensile properties of exposed aliphatic and aromatic TPU films

Sample code	Breaking strength (N/2.5 cm)	Modulus (MPa)	Elongation (%)
TPU-AL	19 ± 1	10.26 ± 1	222 ± 25
TPU-AL-100 h	82 ± 5	10.97 ± 3	700 ± 70
TPU-AL-200 h	62 ± 4	13.47 ± 2	586 ± 10
TPU-AL-300 h	56 ± 2	14.38 ± 3	300 ± 12
TPU-AL-400 h	44 ± 4	16.47 ± 2	235 ± 24
TPU-AL-500 h	11 ± 2	10.85 ± 1	80 ± 4
TPU-AR	20 ± 2	10.1 ± 2	316 ± 30
TPU-AR-100 h	38 ± 1	13.82 ± 5	582 ± 28
TPU-AR-200 h	37 ± 2	14.76 ± 2	589 ± 36
TPU-AR-300 h	5 ± 1	42.38 ± 3	67 ± 31

properties as far as helium permeability is concerned. However, extrapolated data shows that even after 900 h exposure TPU-AL film has helium permeability up to 4.5 L/m^2/24 h.

Conversely, when TPU-AR film was exposed in artificial weathering initially helium barrier increased as exposure time increases up to 300 h which is attributed to the molecular rearrangement that leads to the crosslinked structure in aromatic TPU film, and increasing exposure time gives a higher helium permeability value because the cracks were observed on the surface of aromatic TPU films, these cracks on film surface were also confirmed by SEM analysis. When this data is extrapolated using the polynomial equation, the value of helium permeability is significantly decreased in 900 h which shows permeability up to 30 L/m^2/24 h. Practically, film loses its helium retention property after 900 h. This finding concludes that aromatic film is seriously degraded under UV exposure.

3.3 Surface Degradation

Surface morphology: Surface degradation of TPU films was also assessed by surface morphology and surface topography analysis. SEM analysis was used to study surface roughness and cracks on exposed TPU films (see Fig. 5). SEM micrograph of aliphatic TPU film shows that the surface of the film remain smooth as there were no cracks observed, except slightly roughness on the surface after 500 h. However, SEM micrograph of aromatic TPU film shows cracks after 500 h exposure in artificial weathering, there were clear cracks noticed; however, film still has helium permeability around 1.9 L/m^2/24 h, and it means that these cracks are formed only on the top surface. This finding concludes that aliphatic film is more stable in UV exposure than aromatic film.

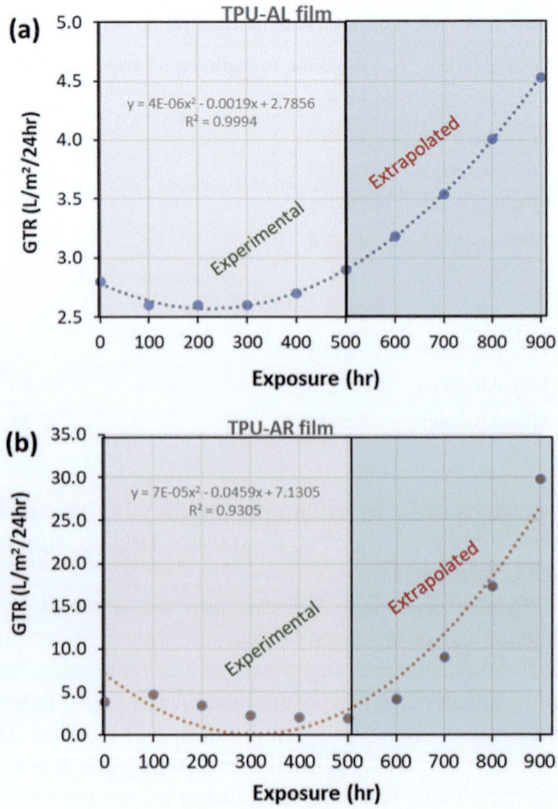

Fig. 4 Helium gas permeability of aliphatic and aromatic TPU film experimental and extrapolated data curve; **a** TPU-AL and **b** TPU-AR films

Table 2 Helium gas permeability of unexposed and exposed TPU films

Samples	Helium gas permeability (L/m^2/24 h)					
	Exposure time					
	0 h	100 h	200 h	300 h	400 h	500 h
TPU-AL	2.8	2.6	2.6	2.6	2.7	2.9
TPU-AR	3.8	4.7	3.4	2.3	2	1.9

Surface topography: Surface roughness was also evaluated by AFM analysis through tapping probe taking a testing area of 20 μm^2. Figure 6 shows the 3D images of AFM analysis after 500 h exposed films of TPU-AR and TPU-AL. Similar to SEM micrographs, AFM images of TPU-AR also show a very high peak height of 61 nm and RMS roughness value of about 68 nm. This is attributed to the noticeable degradation of TPU-AR film during UV exposure in artificial weathering test. On the other hand, TPU-AL film exhibits only roughness at peak height up to 23 nm and RMS roughness around 14 nm. This finding indicates that TPU-Al film remains stable during UV exposure in artificial weathering.

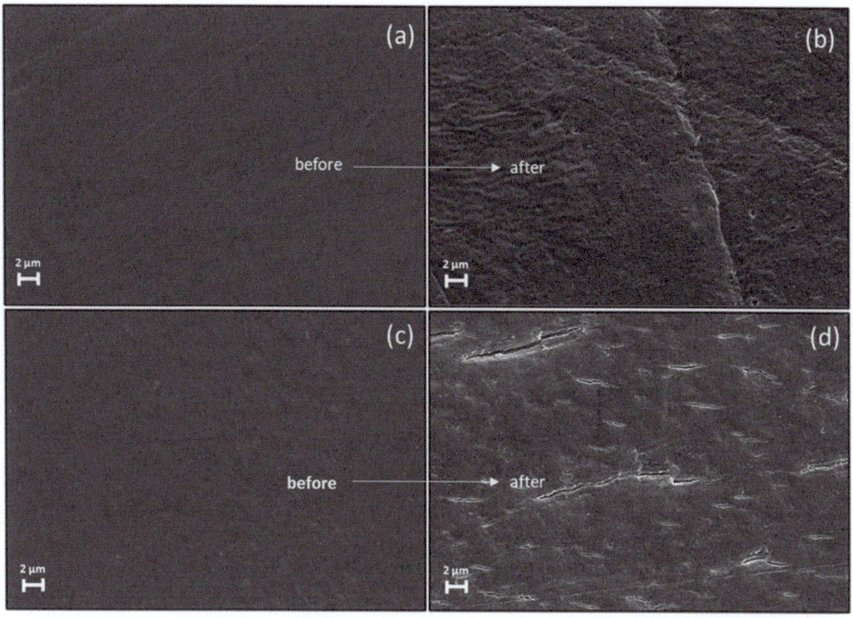

Fig. 5 SEM micrographs of TPU films surface before and after 500 h artificial weathering; **a** and **b** for TPU-AL, **c** and **d** for TPU-AR films surface

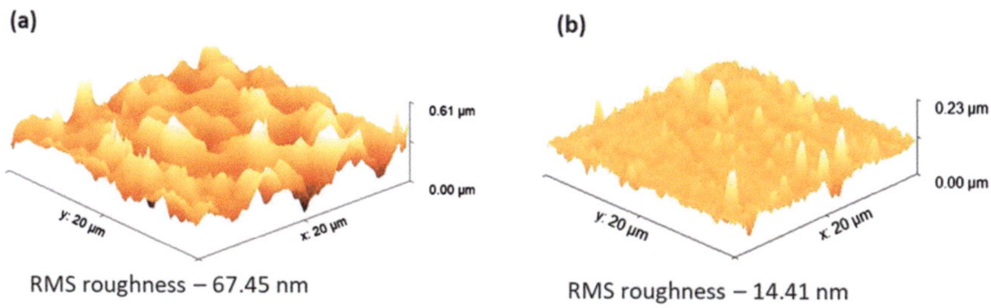

Fig. 6 AFM images of 500 h exposed aliphatic and aromatic TPU films; **a** TPU-AR-500 h and **b** TPU-AL-500 h

3.4 Chemical Structure

ATR-FTIR analysis was used to study the change in chemical structures after artificial weathering. Several characteristic peaks were assigned based on the literature study. Figures 7 and 8 represent the ATR-FTIR spectra of unexposed TPU film, 300 h and 500 h exposed film.

Aliphatic TPU film shows a characteristic peak at 3315 cm^{-1} due to the presence of N–H stretching of urethane linkage. The peak at 1680 cm^{-1} occurs due to the carbonyl stretching of urethane linkage in TPU. Other peaks at 1522, and 1450 cm^{-1}

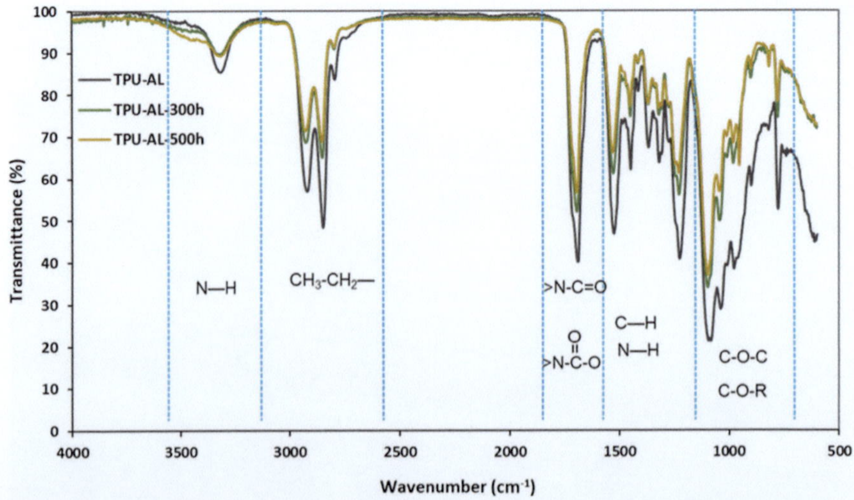

Fig. 7 FTIR analysis of aliphatic TPU films before and after exposure of 300 h and 500 h

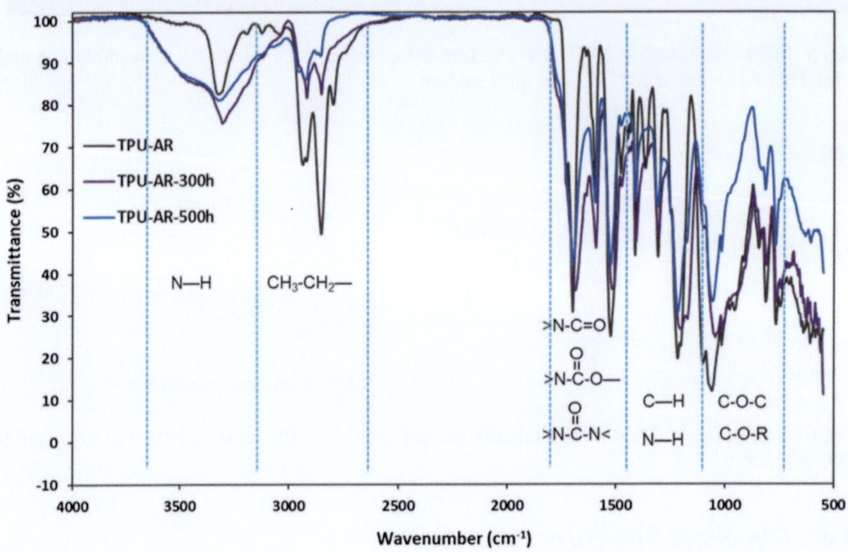

Fig. 8 FTIR analysis of aromatic TPU films before and after exposure of 300 h and 500 h

relates to C-H stretching as well as N–H vibrational peak and peak at 1215 cm^{-1} and 1052 cm^{-1} indicates the presence of C–O–C (ether) groups in TPU films. However, after weathering, TPU-AL film shows a reduction of these peaks due to UV exposure that breaks long molecular chains on the surface of the specimen and forms low molecular weight oxygen containing groups resulting in higher intensity of C-O-peaks between 1250 and 1000 cm^{-1}. Similar characteristic peaks are identified in spectra of aromatic TPU film, peaks at 3315 cm^{-1} N–H stretching and peak at 1680 cm^{-1} $C = O$ stretching due to the urethane linkage of TPU. Other peaks at 1522 and 1412 cm^{-1} relates to C-H stretching and peak at 1215 cm^{-1} and 1052 cm^{-1} indicates the presence of C–O–C groups. As the exposure time interval increases the peak at 1680 cm^{-1} and 3315 cm^{-1} decrease owing to the degradation in TPU-AR film and peak at 1590 cm^{-1} increases due to the formation of urea groups as result of degradation.

3.5 UPF Measurement

The ultraviolet protection factor (UPF) of any material is a measure of the effectiveness of protection against UV radiations coming through the solar spectrum. A higher UPF value of a material indicates low transmission of UV radiation resulting in better protection from degradation by UV. Figure 9 shows the UPF rating of unexposed and exposed aliphatic and aromatic TPU films at different exposure times. The UPF value of both types of TPU films was 52 with unexposed samples. UPF is gradually increased with exposure time interval and reached 140 after 400 h due to structural changes in TPU thereafter reduced to a value of 55 after 500 h exposure due to degradation of molecular chains. Similar behaviour was observed in tensile properties. Conversely, aromatic TPU film shows practically no change up to 200 h exposure, and then it increases up to 310 after 500 h. this increase in UPF could be attributed to the crosslinked TPU structure and yellowing in TPU leading to the partial absorption of UV radiation and resulting in the lowering in the UV transmission.

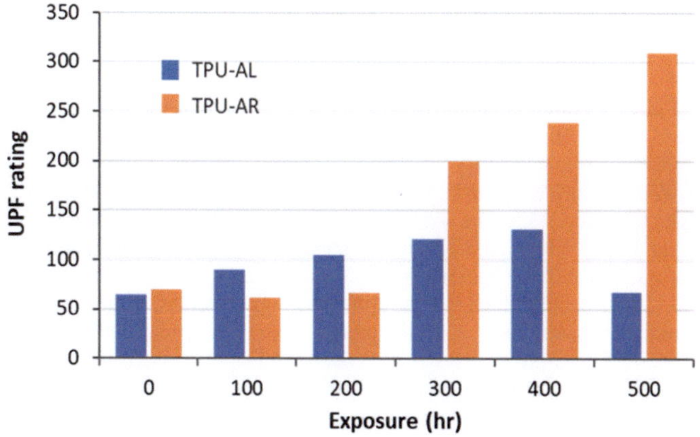

Fig. 9 Change in UV protection value of aromatic and aliphatic TPU films before and after artificial weathering tests

4 Conclusion

TPU-based coating and films are used in designing the envelop structure of LTA systems. The weather stability of these materials is crucial in order to determine the service life of the overall LTA structure. The chemical structure of TPU plays a critical role to determine the weather resistance behaviour. In this work, accelerated artificial weathering is performed to carry out the exposure study. Artificial weathering parameters were specially designed to simulate outdoor natural weathering with a higher exposure rate to UV radiation. Exposure tests carried out on aliphatic and aromatic TPU showed different weather resistance behaviour. With increasing, exposure time TPU film shows improvement in tensile properties at the beginning but further increasing exposure interval leads to the reduction in tensile properties. After 500 h of exposure, both the aliphatic and aromatic TPU film lost the breaking strength of more than 50% than that of unexposed films. Similarly, helium gas permeability initially decreased up to 500 h, and then, films start losing the helium barrier. The aromatic film shows the cracks formation on the surface that led to the loss of the helium barrier. ATR-FTIR analysis also confirms the crosslinked structure formation. This study concludes that aliphatic TPU has more weather resistance properties than aromatic TPU film, which has more reactive groups susceptible to UV degradation during artificial weathering.

Acknowledgements The authors gratefully acknowledge the DRDO collaborating laboratory Aerial Delivery Research and Development Establishment (ADRDE), Agra, India. The authors also acknowledge the Joint Advanced Technology Centre (JATC) a joint venture of DRDO and IIT Delhi for sponsoring this project (RP03465G).

References

1. Zhai H, Euler A (2005) Material challenges for lighter-than-air systems in high altitude applications, AIAA 5th ATIO AIAA 16th Light. Syst. Technol. Conf. Balloon Syst. Conf. 3:1756–1767
2. Scholz P, Wachtendorf V, Panne U, Weidner SM (2019) Degradation of MDI-based polyether and polyester-polyurethanes in various environments - Effects on molecular mass and crosslinking. Polym Test 77:105881. https://doi.org/10.1016/j.polymertesting.2019.04.028
3. Qin J, Jiang J, Tao Y, Zhao S, Zeng W, Shi Y, Lu T, Guo L, Wang S, Zhang X, Jie G, Wang J, Xiao M (2021) Sunlight tracking and concentrating accelerated weathering test applied in weatherability evaluation and service life prediction of polymeric materials: A review. Polym Test 93:106940. https://doi.org/10.1016/J.POLYMERTESTING.2020.106940
4. Ribeiro MCS, Ferreira AJM, Marques AT (2009) Effect of natural and artificial weathering on the long-term flexural performance of polymer mortars. Mech Compos Mater 45:515–526. https://doi.org/10.1007/s11029-009-9104-7
5. Badji C, Beigbeder J, Garay H, Bergeret A, Bénézet JC, Desauziers V (2018) Correlation between artificial and natural weathering of hemp fibers reinforced polypropylene biocomposites. Polym Degrad Stab 148:117–131. https://doi.org/10.1016/J.POLYMDEGRADSTAB.2018.01.002

6. Chatterjee U, Patra S, Butola BS, Joshi M (2017) A systematic approach on service life prediction of a model aerostat envelope. Polym Test 60:18–29. https://doi.org/10.1016/j.polymertesting.2016.10.004
7. U. ATLAS Inc. (2001) Weathering testing guidebook, ATLAS. Weather Serv Gr, 1–108
8. Quill J (2001) The essentials of laboratory weathering Q-Lab corporation. Q-Lab Corp, 1–16
9. Liu Y, Liu S, Huifeng T (n.d.) Effect of accelerated Xenon lamp aging on the mechanical properties and structure of thermoplastic polyurethane for stratospheric airship envelope. Sci Technol Adv Compos Spec. https://doi.org/10.1007/s11595-014-1080-7
10. Boubakri A, Guermazi N, Elleuch K, Ayedi HF (2010) Study of UV-aging of thermoplastic polyurethane material. Mater Sci Eng A 527:1649–1654. https://doi.org/10.1016/j.msea.2010.01.014
11. Boubakri A, Haddar N, Elleuch K, Bienvenu Y (2010) Impact of aging conditions on mechanical properties of thermoplastic polyurethane. Mater Des 31:4194–4201. https://doi.org/10.1016/J.MATDES.2010.04.023

A Multidisciplinary Design Optimisation (MDO) Algorithm for the Automatic Sizing of an Unmanned Lighter-Than-Air Platform

Piero Gili, Ludovica Castronovo, Marco Civera, Rinto Roy, and Cecilia Surace

1 Introduction

This study presents a multidisciplinary design optimisation (MDO) algorithm, conceptualised and developed for the automatic design and sizing of an unmanned lighter-than-air platform (LTA).

The MDO method was chosen to solve the design of the airship since the nature of the problem is based on the integration of several systems and the need to think of an optimisation for the sizing of the entire airship, considering each of the systems and their relations.

This unmanned aerial system (UAS), presented in [1] and further detailed in [2] and [3], is intended for remote sensing at relatively low altitudes, aiming at detailed land use and land cover (LULC) mapping at urban or rural level (i.e. for areas between 10^5 and 10^7 m^2). The proposed MDO approach is intended to automatically return the best design and sizing according to the intended mission requirements, simplifying the design process. Overall, the proposed algorithm shows the capability to automatically size the structure for the conditions of interest, as well as to return the preferable energy system, choosing between two options (fuel cells or batteries).

Further studies are developed as a sensitivity analysis, to test the possibility to use the algorithm for new missions, with new parameters; to do so, the analysis is focused on the response of each subsystem of the platform (in terms of weight) to the variation of a certain parameter.

P. Gili (✉) · L. Castronovo
Department of Mechanical and Aerospace Engineering, Politecnico di Torino, Turin, Italy
e-mail: piero.gili@polito.it

M. Civera · R. Roy · C. Surace
Department of Structural, Geotechnical and Building Engineering, Politecnico di Torino, Turin, Italy

P. Gili · M. Civera · R. Roy · C. Surace
Interdepartmental Responsible Risk Resilience Centre (R3C), Politecnico di Torino, Turin, Italy

D. Shukla (ed.), *Lighter Than Air Systems*, Lecture Notes in Mechanical Engineering,
https://doi.org/10.1007/978-981-19-6049-9_10

1.1 Problem Definition and Formulation

LTA platforms can serve a well-defined niche for earth observation and remote sensing. Specifically, they can be deployed at a lower cost than other heavier-than-air fixed- or rotating-wing alternatives to cover small to large areas (Figs. 1 and 2). More specifically, an unmanned remotely piloted LTA can satisfy different applications for LULC mapping and environmental surveillance. The specific unconventional LTA discussed here (Fig. 3) further integrates all these advantages with an innovative propulsion and control system, without aerodynamic control surfaces, a thrust vectoring technology [4, 5] with a dedicated control law [6], and an axi-symmetric, ellipsoidal envelope. These advantages are graphically summarised in Fig. 4.

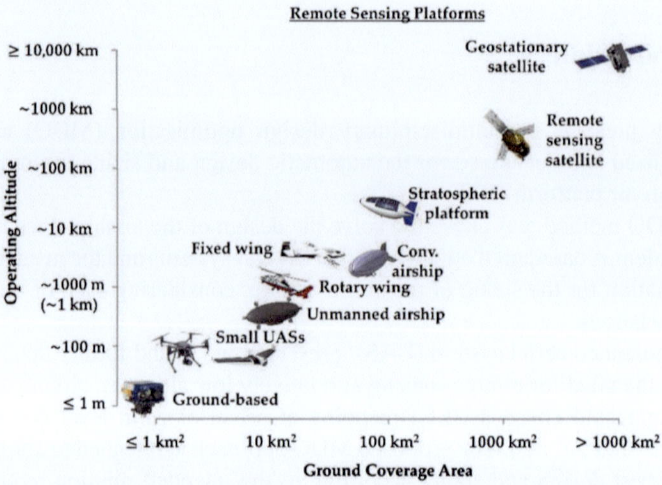

Fig. 1 Comparison of remote sensing platforms. *Source* Retrieved from [1]

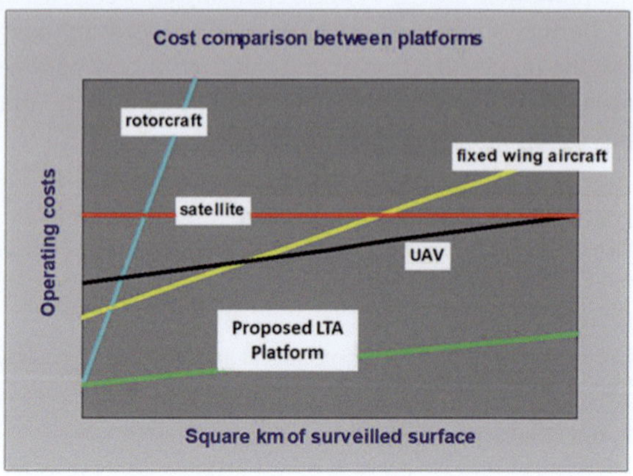

Fig. 2 Qualitative estimates of costs per surveyed square kilometre. *Source* Retrieved from [1]

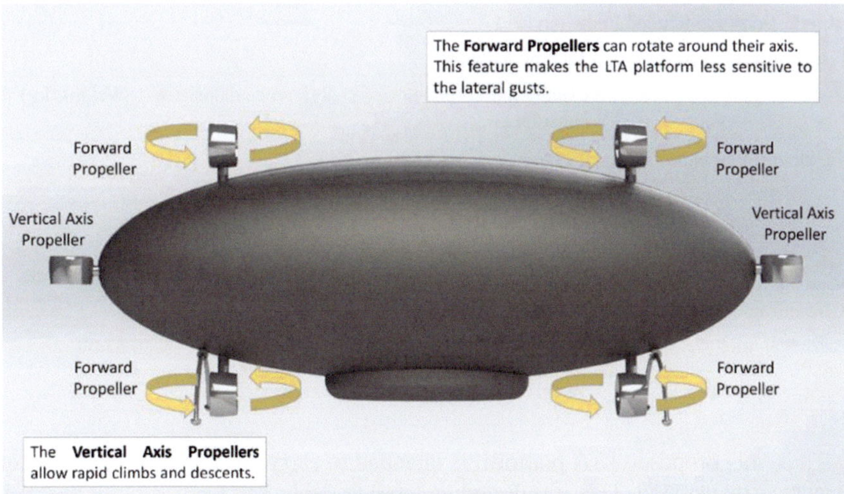

Fig. 3 3D rendering of the LTA platform intended for this study, with an illustration of the propulsion system. *Source* Retrieved from [1]

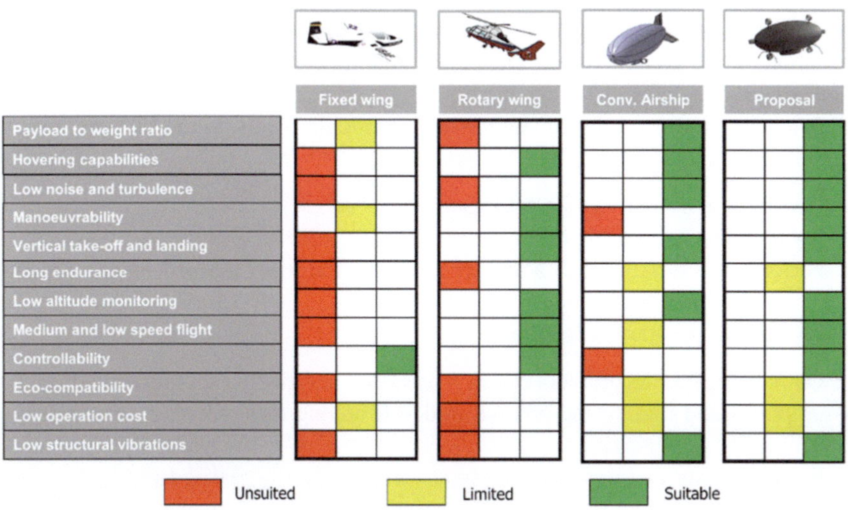

Fig. 4 Advantages and limitations of comparable land survey strategies. *Source* Retrieved from [1]

This specific design derives from previous studies, performed throughout the last twenty years [7–10] and that led to the proposal described in [1]. From a structural perspective, it is a single-rib design, with the six propellers and the payload gondola directly attached to this sole load-carrying element.

Table 1 Sensors included in the payload

Target measurement	Technical details		
	Commercial product considered	Reference datasheet	Weight (kg)
Visible infrared bands	1 Phase one® iXM-RS150F camera	[12]	2.15
Thermal infrared bands	1 NEC® TH9260 thermographic camera	[13]	1.70
Near, short wave, and mid wave infrared bands	1 SPECIM® Aisa DUAL hyperspectral system	[14]	50.00
Laser scanning	1 Optech® ALTM Galaxy T1000 laser scanner	[15]	33.50

Thus, this proposed LTA platform is intended to carry a relatively large array of sensors as its payload. This is reported in detail in Table 1.

However, even having the payload and the control and propulsion system defined, both the project and the sizing of the remaining components are not trivial. For instance, there is no a priori optimal choice for the power system; both fuel cells or batteries could be potentially preferable, depending on the mission requirements (in particular, the survey area and the distance from the mission location). In turn, the selection of one specific option affects the total mass of the system, hence the final size (length and diameter) of the airship.

All these parameters depend on the mission requirements, as the LTA platform is supposed to:

- Depart from a suitable take-off location;
- Fly to the target survey area at a given cruise speed and altitude;
- Perform the survey (with a survey speed and altitude defined by the sensing devices' limitations);
- Fly back to the starting location.

One can see that the problem can only be solved iteratively, incorporating different assumptions and finding the optimal trade-off among different sizes and different power supply strategies.

For this reason, an optimisation code was specifically realised in MATLAB to automatise the design process. This follows the main principles of multidisciplinary design optimisation [11].

2 Methodology

In general, MDO is a method to solve design problems that include different disciplines, by optimising simultaneously the entire system, not every discipline separately; it is therefore essential to analyse and use the relations and interactions between the subsystems in order to get an overall understanding of the problem.

The main steps to develop a MDO algorithm are: (1) the selection of the design variables and their constraints (in this case, some mission parameters and the respective ranges of interest), (2) the selection of the objective to minimise or maximise (in the design of an airship, this is represented by the weight or volume which should be minimised), and (3) a model to link these objects.

Several MDO strategies are already available from the published scientific literature (see e.g. [16]). However, almost all of these algorithms and concepts are intended for heavier-than-air platforms, which have different needs and limitations than the ones of interest here.

Thus, a novel strategy, not based on any existing alternatives, is developed. The code consists of an iterative procedure varying four main parameters:

1. The total length of the LTA platform L;
2. The cruise speed (from the take-off point to the target survey area) v;
3. The cruise altitude (from the take-off point to the target survey area) z;
4. The slenderness ratio of the envelope d/L.

These are varied in pre-set intervals of interest. For each iteration, a logical process is followed that leads to the selection of only those solutions that meet the mission requirements and sustenance of the airship; finally, among these, the combination of parameters that guarantees the lowest weight is saved and stored.

The script, therefore, allows obtaining the size and weight of the airship as an output, especially considering how the masses are distributed (in percentage) among the several components (payload, load-bearing structure, power system, etc.).

The code is intended to adapt for the specific mission requirements, as described before, receiving the following inputs:

1. Survey speed, in m/s;
2. Survey altitude, in m;
3. Payload, in kg (considering the one reported in Table 1, a subset of it, or another selection of remote sensing devices);
4. Survey area, in m^2.

The strategy followed is one of concurrent subspace optimisations [17]. That is to say, three subsystems—the power system, the propulsion system, and the load-bearing structure— are concurrently optimised. This is necessary since these subsystems are coupled with one another.

All the components of the energy and propulsive systems are computed for both the cruise and the survey phases, in terms of mass and associated power. Eventually, the higher weight solution is saved; this step is summarised in Fig. 5 under the label 'on-board system', and displayed at the side of the complete diagram.

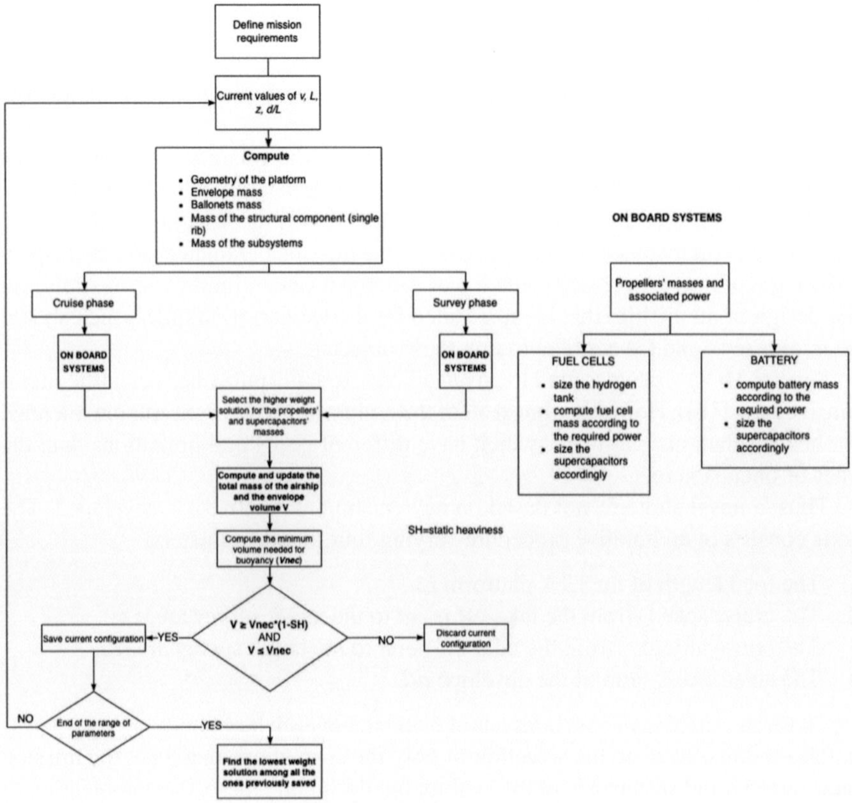

Fig. 5 Flow chart of the complete MDO algorithm

Each solution, if feasible, is saved. At the end of the process (i.e. after all the four main parameters have been tested on their respective ranges, considering all the potential combinations), the feasible solutions are searched for the optimal trade-off. The complete algorithm is graphically depicted in Fig. 5.

3 Results

3.1 Variation of Payload and Associated Power

Several combinations of sensors have been evaluated, with and without the laser scanning and/or the hyperspectral camera. The removal of these sensors, which contribute the most to the estimation of the payload and the related power, has a direct influence on the total mass of the airship, affecting also the energy system; however, it does not involve relevant changes in the optimal configuration of the platform, and it still ensures the advantage of the fuel cells system above the use of batteries.

3.2 Energy Systems

A comparison between the two proposed energy systems has been developed, to establish which solution is more convenient, in terms of weight, as a function of the distance from the mission location; this analysis has been repeated for different values of the survey area (10^5–10^6–10^7 m^2).

The outcomes of this specific study show that in case of low autonomies (i.e. for a survey area of 10^5 m^2), the battery configuration proves to be more convenient below a certain distance from the mission location. The results are displayed in Figs. 6 and 7 as graphs that illustrate the relationship between the distance and the respective weight of the two energy systems.

Furthermore, a methodic sensitivity analysis is performed accordingly to the mission requirements—considering a cruise altitude varying between 500 and 1000 m, a cruise speed between 10 and 30 m/s, a survey area between 10^5 m^2 and 10^7 m^2, and a length of the airship fixed between 30 and 50 m.

Out of all the feasible solutions, the fuel cell solution prevailed over the battery option for all the possible mission requirements except that for very small areas and very short cruise distances (i.e. when the required autonomy was minimal), showing a result that proves to be consistent with the outcomes of the original algorithm for the design of the airship. It was found that the cruise altitude affects the overall mass especially due to the weight increase of the ballonets. The cruise speed has an

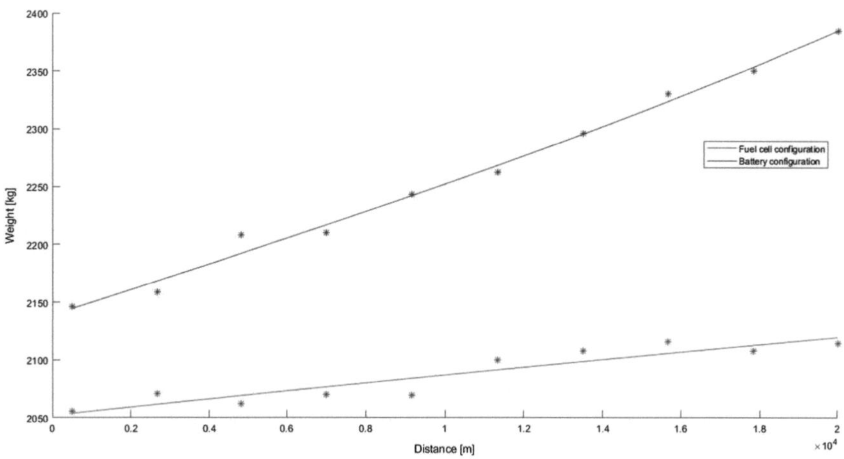

Fig. 6 Airship weight as a function of the distance from mission location, for the two energy systems, considering a survey area of 10^7 m^2

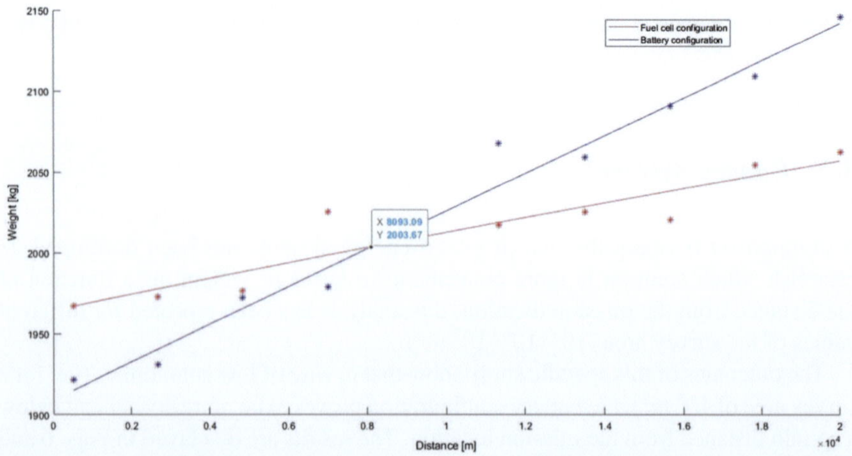

Fig. 7 Airship weight as a function of the distance from mission location, for the two energy systems, considering a survey area of 10^5 m^2

impact on flight time and the power needed when cruising: higher speeds require more energy, thus heavier solutions, which must provide more power to overcome the increased drag. Nevertheless, at least one feasible solution was found for any specific set of mission requirements. Indeed, under all conditions, the automatic MDO algorithm returned designs that were deemed acceptable from an expert user.

4 Discussion and Conclusions

The topic of multidisciplinary design optimisation is of great interest in the field of aerospace design, yet not much addressed for the specific case of LTA platforms, let alone for RPAS medium- to large-size airships. The proposed algorithm allows sizing and designing an unmanned, remotely piloted, low-altitude LTA in a parallel and automated fashion, accordingly to the mission requirements. This automation can return the optimal LTA parameters almost immediately. This is a noteworthy improvement with respect to the long time required by human-made, sequential, trial-and-error-based traditional design procedures. For this specific aim, the goal (at both system and subsystem level) was to find the feasible design which satisfies all the constraints for the minimum total mass and given survey conditions. These constraints were mainly weight and power balance, that is to say, to guarantee buoyancy and self-propelling capabilities at any time throughout the whole duration of the mission. However, this can be further modified if needed, making the algorithm even more versatile and applicable for similar tasks, not limited to LULC survey.

References

1. Gili P, Civera M, Roy R, Surace C (2021) An unmanned lighter-than-air platform for large scale land monitoring. Remote Sens 13:2523
2. Gili P, Civera M, Roy R, Cibrario Bertolotti S, Cosenza D, Gili A, Surace C (2021) Design of a prototype unmanned lighter-than-air platform for remote sensing: control, alimentation, and propulsion systems. In: Proceedings of the 32nd Congress of the International Council of the Aeronautical Sciences (ICAS 2021), Shanghai
3. Surace C, Roy R, Civera M, Allaio D, Barbieri R, Grava A, Tufilli R, Gili P (2021) Design of a prototype unmanned lighter-than-air platform for remote sensing: structural design and optimization. In: Proceedings of the 32nd Congress of the International Council of the Aeronautical Sciences (ICAS 2021), Shanghai
4. Gili P, Lerro A, Vazzola M, Visone M (2011) A new approach for the estimation of the aerodynamic damping characteristics of the ETF demonstrator. SAE Int J Aerosp 4:1115–1124
5. Battipede M, Lando M, Battipede M, Lando M, Gili PA (2004) Peculiar performance of a new lighter-than-air platform for monitoring
6. Battipede M, Lando M, Gili P (2006) Mathematical modelling of an innovative unmanned airship for its control law design. In: Proceedings of the IFIP International Federation for Information Processing. Springer, Boston, MA, Vol 202, pp 31–42
7. Battipede M, Gili P, Maggiore P, Lando M (2006) Risk assessment and failure analysis for an innovative remotely-piloted airship. In: Proceedings of the Collection of Technical Papers - 6th AIAA Aviation Technology, Integration, and Operations Conference, Vol 1, pp 119–129
8. Cappadona A, Lecca R, Vazzola M, Gili P, Farina P, Surace C (2009) Innovative unmanned airship structural analysis: Dual-hull and exoskeletal configurations. J Phys: Conf Series Institute of Physics Publishing 181:12097
9. Battipede M, Gili P, Lando M (2005) Ground station and flight simulator for a remotely-piloted non conventional airship. In Proceedings of the Collection of Technical Papers—AIAA Guidance, Navigation, and Control Conference 5:3660–3671
10. Battipede M, Gili P, Vazzola M (2013) Structural and aerodynamics analysis on different architectures for the elettra twin flyer prototype. J Intell Robot Syst Theory Appl 72:123–144
11. Sobieszczanski-Sobieski J (1993) Multidisciplinary design optimization: An emerging new engineering discipline
12. iXM-RS150F technical data sheet. https://downloads.phaseone.com/7e71b0a2-f42d-42c7-896b-2c16968f79c3/English/iXM-RS_Brochure_letter.pdf. Last accessed 8th march 2021
13. Thermo Tracer TH9260 technical data sheet. https://www.infrared.avio.co.jp/en/products/ir-thermo/lineup/th9260/index.html. Last Accessed 8th March 2021
14. AISA DUAL hyperspectral sensor technical data sheet. http://www.adept.net.au/cameras/specim/systems/pdf/Aisa_Dual.pdf. Last accessed 8th March 2021
15. ALTM Galaxy T1000 Airborne Lidar Terrain Mapper technical data sheet. https://geo3d.hr/sites/default/files/2018-06/2017-11-02_Optech_Galaxy-Brochure_HR.pdf. Last accessed 8th March 2021. Available online: www.teledyneoptech.com. Accessed on 8th March 2021
16. Martins JRRA, Lambe AB (2013) Multidisciplinary design optimization: a survey of architectures 51:2049–2075. https://doi.org/10.2514/1.J051895
17. Liang H, Zhu M, Guo X, Zheng Z (2012) Conceptual design optimization of high altitude airship in concurrent subspace optimization. 50th AIAA Aerosp Sci Meet Incl New Horizons Forum Aerosp Expo, 1–17

The Concepts of Telescopic and Self-Deployable Tensegrity-Based Helium-Filled Aerostats

Lech Knap, Andrzej Świercz, Cezary Graczykowski, and Jan Holnicki-Szulc

1 Introduction

Airships and balloons were the first flying vehicles but for a long time, their usage has been very limited due to development of aviation. Specified military application included reconnaissance missions and ocean surveillance [1, 2], while the civil use was interrupted by crash of the Hindenburg in 1937. Recent technological changes have caused renaissance of airship applications. At the beginning of the century, many new-generation helium-filled airships enabling long-term and low-energy flights have been constructed and used for low-cost stratosphere and mesosphere exploration [3, 4]. Currently constructed airships provide telecommunication in rarely populated areas, serve as remote monitoring systems and research pseudo-satellites [5–7]. The most well-known examples of airships are Zephyr S, Russian concept Berkut, Lockheed Martin's ISIS and French Stratobus.

The first part of this contribution proposes adaptive telescopic aerostat [8] with multi-segmented construction, which is equipped with controllable segments' couplings, additional pressure tank with compressor and valve. The proposed system allows to modify pressure difference between the aerostat and the atmosphere, and to obtain desired volume of the aerostat at each altitude. As a result, the vertical

L. Knap
Institute of Vehicles and Construction Machinery, Warsaw University of Technology, Warsaw, Poland
e-mail: lech.knap@pw.edu.pl

A. Świercz · C. Graczykowski (✉) · J. Holnicki-Szulc
Institute of Fundamental Technological Research, Polish Academy of Sciences, Warsaw, Poland
e-mail: cgraczyk@ippt.pan.pl

A. Świercz
e-mail: aswiercz@ippt.pan.pl

J. Holnicki-Szulc
e-mail: holnicki@ippt.pan.pl

D. Shukla (ed.), *Lighter Than Air Systems*, Lecture Notes in Mechanical Engineering,
https://doi.org/10.1007/978-981-19-6049-9_11

motion of the aerostat can be controlled in arbitrarily selected manner. In particular, flight path with pre-defined altitude changes can be obtained with minimal transfer times or with minimal energy requirements for the compressor. Two corresponding control strategies aimed at maximal control efficiency and minimal control cost are presented and compared in the following part of the paper.

In the second part of this contribution, the authors propose aerostat based on deployable tensegrity structure with active elements of controllable lengths [9, 10]. Such construction enables convenient transport of the aerostat in a compact form with the use of an hot-air balloon or an aircraft and its automatic deployment at operational altitude. The tensegrity structure provides low mass of the aerostat, adequate support of the aerostat envelope and possibility of transferring point loads resulting from carried cargo. On the other hand, it enables convenient change of the aerostat volume and shape during the flight. The change of volume can be used to control the actual buoyancy force and resulting vertical motion of the aerostat ("V-mobility"). In turn, the change of aerostat shape can be used for changing the influence of lateral wind gusts or controlling horizontal stability during the mission ("H-stability"). In both above cases, the attention of conducted research is focused on strength analysis during the process of morphing.

2 The Concept of Adaptive Telescopic Aerostats

The proposed concept for vertical mobility is based on the reversible, telescopic option for the aerostat volume modifications, in which the crucial feature is system composed of structural sections connected via controllable actuators fixing or releasing their relative movement and additional helium container (pressure tank) allowing controllable release or absorption of gas [8]. The design concept of the small high-altitude pseudo-satellite (HAPS) module applied in further analyses of its mobility and load-carrying capacity is illustrated in Figs. 1 and 2. The construction consists of a cylindrical housing reinforced by axial and circumferential ribs.

Fig. 1 Scheme of the HAPS design: aerostat in compact form

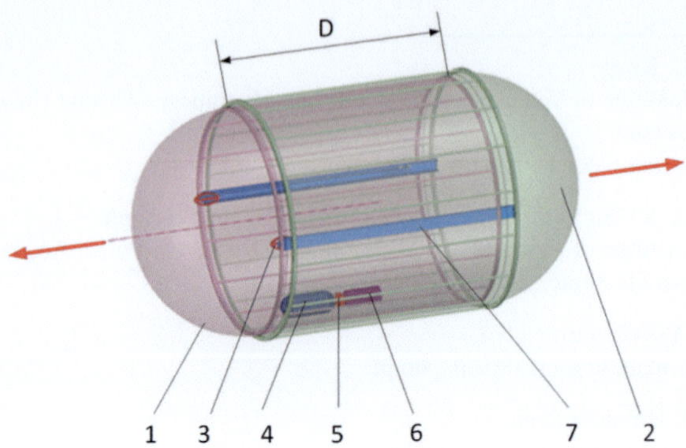

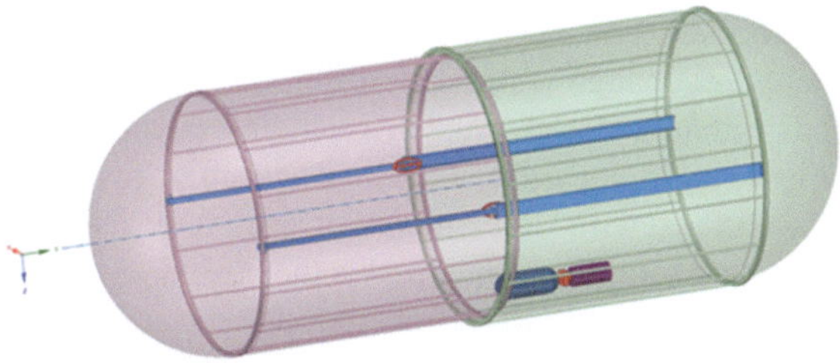

Fig. 2 Scheme of the HAPS design: aerostat in extended form

The HAPS offers a possibility of volume modification by changing the length of the cylindrical section from initial dimension of D to the maximal dimension of 2D. It is possible due to the construction consisting of the inner (1) and outer segments (2), which are sliding along linear guideways (7). Relative motion of these guideways can be blocked by actuator (3), which creates the clamping mechanism providing control of the aerostat volume.

In order to achieve a proper mobility, the aerostat is equipped with an additional gas storage tank (4) with an additional amount of pressurized helium. This additional tank serves for the purpose of generating the pressure difference between aerostat and its environment. The created overpressure or underpresure is used to generate forces needed to fold and unfold the aerostat. In order to reduce helium losses caused by its release to atmosphere, the aerostat is equipped with a compressor (6) and a valve (5), which control the flow of helium from the aerostat back to the additional gas tank.

Application of the above described system composed of additional pressure tank, the compressor and the valve enables precise control of the gas flow and actual pressure difference between the aerostat and the atmosphere during the ascending and descending processes. This allows to control the process of folding and unfolding of the aerostat, to generate the required value of the lift force resulting from buoyancy and to obtain equilibrium position of the aerostat at desired altitude.

3 Comparison of Strategies for Control of Aerostat Motion

In the considered control problem, we will analyse the process of aerostat ascending and descending of total duration of 12,000 s. During the first part of the process, the aerostat should ascend from the ground level to initial equilibrium position located at $h_1 = 350$ m, next it should float to the target altitude $h_3 = 5000$ m with an intermediate stop at the altitude $h_2 = 3000$ m. Further, in the second part of the process, the aerostat should descend from the altitude $h_3 = 5000$ m back to altitude $h_5 = 350$ m, with the intermediate stop at the altitude of $h_4 = 3000$ m.

The first control problem is to find change of aerostat volume and the corresponding required change of pressure difference between the aerostat and the atmosphere, which enable realization of the above defined aerostat flight path, providing the shortest time motion between the assume altitudes. The solution involves maximally fast change of pressure difference and step-wise change of aerostat volume at the beginning of each stage of ascending and descending. The conducted simulations (Figs. 3 and 4, *red lines*) indicate that each stage of ascending and descending process is associated with large amount of work done by the compressor. In particular, large power consumption is required at the beginning of the descending stages, when large mass of helium has to be transferred to the storage tank to reduce the volume of the capsule and allow atmospheric pressure to fold the structure.

The second control problem is to find change of aerostat volume and the corresponding required change of pressure difference between gas pressure in the aerostat and the atmospheric pressure, which enable realization of the above defined aerostat flight path, providing minimal energy used for gas transfer between additional gas tank and aerostat. The optimal solution leads to realization of the entire process in possibly slow manner, with minimal pressure difference between gas pressure in the aerostat and gas pressure in the storage tank. In the conducted simulations

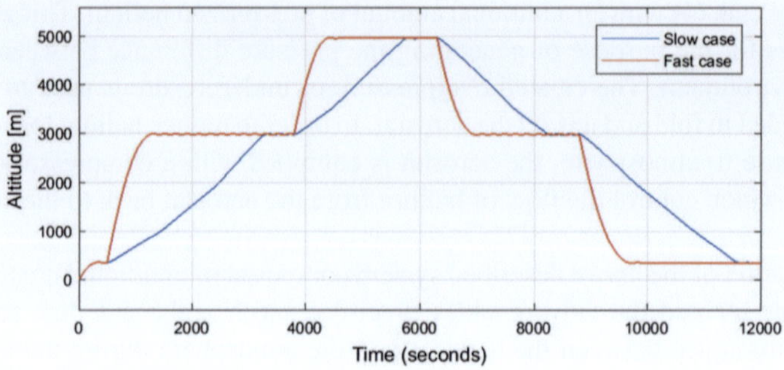

Fig. 3 Change of aerostat altitude: strategy maximizing vertical mobility efficiency (*red*) strategy minimizing vertical mobility energetic cost (*blue*)

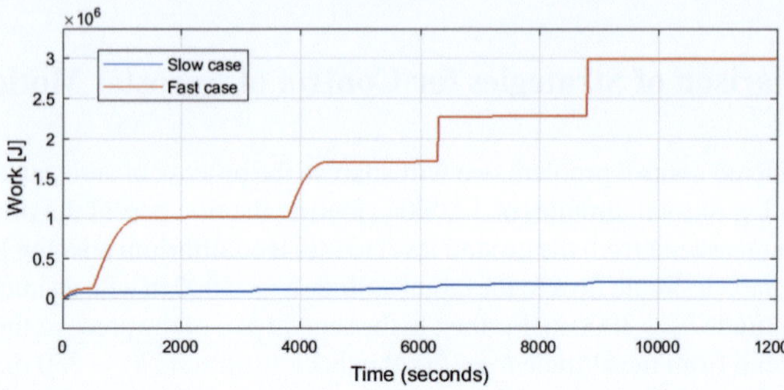

Fig. 4 Work during gas transfer: change of aerostat altitude: strategy maximizing vertical mobility efficiency (*red*), strategy minimizing vertical mobility energetic cost (*blue*)

(Figs. 3 and 4, *blue lines*), the modification of aerostat volume is conducted progressively during subsequent pre-defined time periods of ascending and descending. As a result, each stage of the process is associated with significantly smaller work done by compressor. In particular, initial parts of descending stages do not require excessive power consumption.

4 The Concept of Tensegrity-Based Aerostat

The simplest tensegrity-based aerostat utilizes "tensegrity prism" composed of three struts and nine tendons as an internal skeletal structure (Fig. 5a). Selected nodes (1a) of the tensegrity are permanently integrated with aerostat envelope, while the other nodes (1b) are connected slidably in order to avoid excessive internal forces. The applied tensegrity structure differs from the classical one since three vertical tendons (*green-coloured*, 1e) are highly elastic, while the horizontal stiff tendons (*blue-coloured*, 1c) have controllable lengths. Structural stability of the structure is ensured by the stiff struts (*red-coloured*, 1d).

The aerostat based on such a tensegrity structure has cylindrical shape with two hemispheres located at the lateral sides (Fig. 5b). Symmetric shortening of the stiff tendons of controllable lengths implies change of cylinder radius, decease of aerostat volume and causes that aerostat shape becomes more slender. In turn, asymmetric shortening of the tendons causes asymmetric deformation of the aerostat and changes its resistance to winds in axial direction.

The more complex tensegrity-based aerostat has a skeletal structure composed of four tensegrity modules. Each tensegrity unit is based on two squares rotated by 45° (Fig. 1c–d), in which the stiff tendons (1c) are arranged into crosses. The struts, highly elastic and controllable tendons are shown in Fig. 5c. The set of movable joints of each unit tensegrity structure is slidingly connected to guide rails, which are bounded to the aerostat envelope.

In this case, the aerostat has a shape of ellipsoid with a circular cross section in the vertical plane. The considered aerostat is 10 m long and 2 m wide, and it is divided into three chambers with the use of two internal diaphragms. Application of four tensegrity modules provides wide possibilities of changing aerostat shape. In the proposed solution, the shape of the arbitrary section can be significantly modified, which causes that aerostat can be precisely adjusted to wind gusts in both axial and lateral directions, leading to the process of "adaptive morphing." Such process can be efficiently conducted using controllable retractors, such as rotary electric motors or linear actuators, serving for contraction or elongation of stiff tendons.

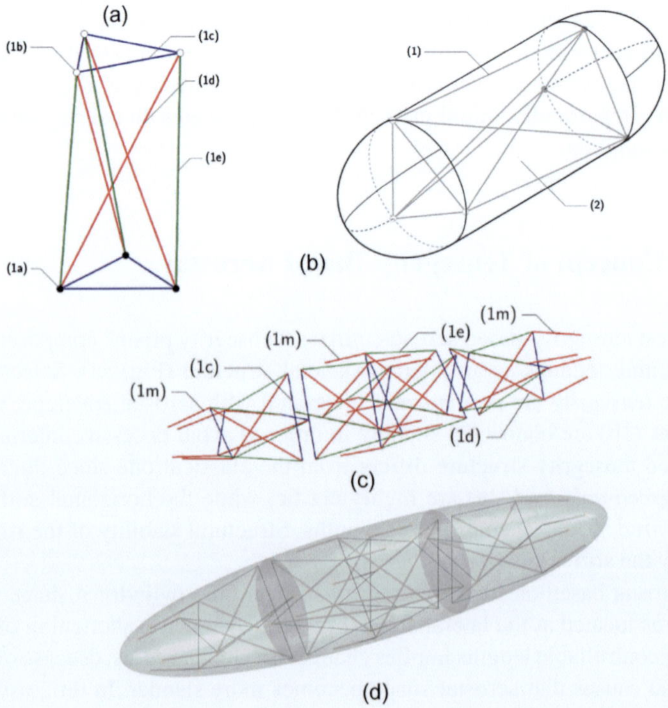

Fig. 5 **a** Single module of foldable tensegrity structure; **b** Single-chamber tensegrity aerostat; **c** Four-module tensegrity structure; **d** Three-chamber aerostat with four modules of tensegrity structure

5 Strength Analysis of Tensegrity-Based Aerostat

The FEM-based strength analysis conducted in ABAQUS software was aimed at finding distributions of internal forces in skeletal structure and aerostat envelope caused by pressure differences at various altitudes. The mesh of the envelope and nodes of the tensegrity structure is illustrated in Fig. 6. The envelope and diaphragms are modelled by means of 3- and 4-node membrane elements, and the stiffening plates in the locations of connection with tensegrity joints and guide rails—by shell elements, bars and guide rails—by beam elements, highly elastic tendons—by truss elements, stiff tendons, retractors and sliding connections—by special type of elements called connectors. For computations, the following material data was assumed:

- envelope and diaphragms: $t = 50\ \mu\text{m}$, $\rho = 1390\ \text{kg/m}^3$ and $E = 0.43$ GPa;
- bars, guide rails (pipe cross-section, $t = 1$ mm) and stiffening plates ($t = 1$ mm)—carbon fibre: $\rho = 1580\ \text{kg/m}^3$ and $E = 87.0$ GPa;
- highly elastic tendons (rubber-like material): $\rho = 900\ \text{kg/m}^3$ and $E = 0.05$ GPa;
- stiff tendons: $t = 3$ mm, $\rho = 7850\ \text{kg/m}^3$ and $E = 210$ GPa.).

Fig. 6 Numerical model of a three-chamber aerostat with four tensegrity modules

The total mass of the aerostat including payload equals to 15.2 kg. A structural strength assessment was performed for three altitudes: $h_1 = 5$ km, $h_2 = 3.725$ km and $h_3 = 2$ km. Firstly, to reach the altitude of h_1, helium mass of 2.537 kg was inflated into the aerostat. Then, aerostat was partially deflated by transferring 0.095 kg of helium to additional tank. Finally, all stiff tendons were uniformly shortened to decrease aerostat volume and achieve the altitude of $h_3 = 2$ km. The changes of atmospheric pressure were based on NASA's atmosphere model.

The equivalent von Mises stresses in the envelope computed for all considered altitudes are presented in Fig. 7. The highest level of the stresses is reached for the altitude of $h_1 = 5$ km—about 50 MPa in the membrane and 67 MPa in the stiffening plates. Descents to the altitude $h_2 = 3.725$ km, being the results of helium transfer, causes that equivalent von Mises stresses are decreasing. When the aerostat continues its descent due to tendons shortening, the stresses in the bar members of tensegrity structure sharply increase to 240 MPa, but they do not exceed the yield stresses limit of the applied material.

6 Discussion and Conclusions

The paper presents positive preliminary verification of two concepts of adaptive aerostats: (i) telescopic aerostats with multi-segmented construction and (ii) aerostats based on self-deployable tensegrity structures. In the case of telescopic aerostats, the conducted simulations prove that control of internal pressure and resulting extension of the aerostat can be used to achieve desired altitudes possibly fast and provide significant reduction of energy consumption. In the case of tensegrity-based aerostats, the aerostat volume and resulting equilibrium altitude can be effectively modified by control of tendons, and such operation does not induce excessive values of stresses. It can be concluded that both proposed constructions significantly increase controllability of aerostat vertical mobility, and thus they are promising solutions in future applications.

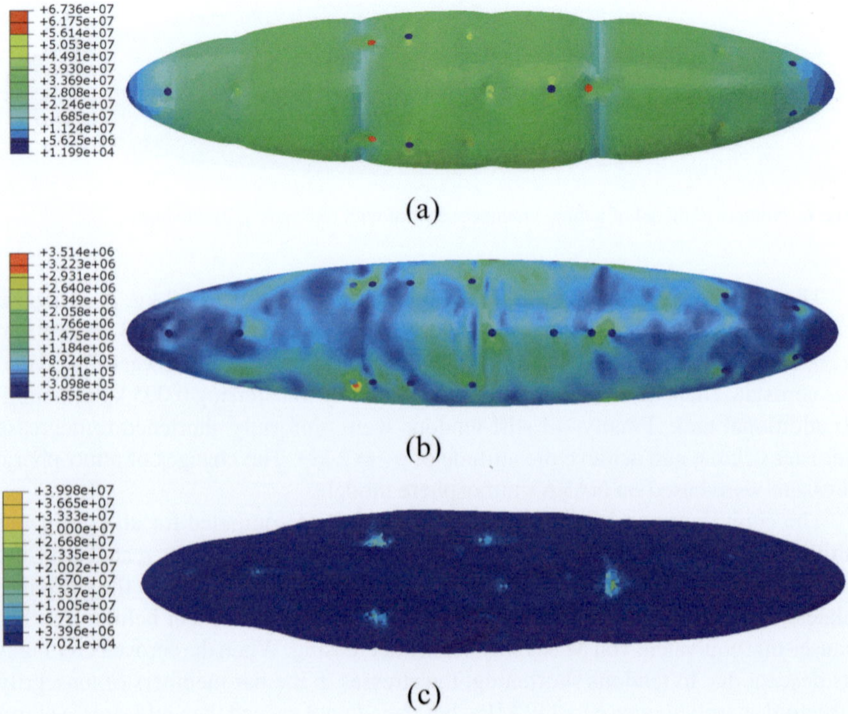

Fig. 7 Equivalent von Misses stresses in the envelope at altitude of **a** $h_1 = 5$ km, **b** $h_2 = 3.725$ km and **c** $h_3 = 2$ km

Acknowledgements The authors acknowledge the support of the National Centre for Research and Development and the National Science Centre, Poland, granted in the framework of the TANGO 4 programme (project TANGO-IV-C/0001/2019-00).

References

1. Brayan T, Rava G (2009) British airships 1905–1930. Osprey Publishing, New York
2. Ege L (1974) Baloons and airships. Blandford Press, London
3. Ghanmi A, Sokri A, Airships for military logistics heavy lift. Defence R&D Canada
4. Future Aerostat and Airship Investment Decisions Drive Oversight and Coordination Needs, Report to the Subcommittee on Emerging Threats and Capabilities, Committee on Armed Services, U.S. Senate, GAO, 2012
5. Saleh S, Weiliang HE (2018) New design simulation for a high-altitude dual-balloon system to extend lifetime and improve floating performance. Chin J Aeronaut 31(5):1109–1118
6. Araripe d'Oliveira F, Lourenço de Melo FC, Devezas TC (2016) High-altitude platforms—present situation and technology trends. J Aerosp Technol Manag So José dos Campos 8(3):249–262
7. Eguchi K, Yokomaku Y, Mori M (2000) Overview of stratospheric platform airship R&D program in Japan. In: AIAA 14th LTA TCCE, Acron, OH, July

8. Knap L, Graczykowski C, Holnicki-Szulc J, Wołejsza Z (2020) Strategies for reduction of energy consumption during ascending and descending process of modern telescopic HAPS aerostats. Bull Pol Acad Sci: Techn Sci. https://doi.org/10.24425/bpasts.2020.131833,68(1),pp. 155-168
9. Holnicki-Szulc J, Świercz A, Kostro S, Knap L, Graczykowski C (2021) A concept of the SDT (Self-Deployable Tensegrity) structure for the rapid and precise lifting of helium aerostats, especially into the stratosphere, Patent No. 3770352, EPO, 2021
10. Knap L, Świercz A, Graczykowski C, Holnicki-Szulc J (2021) Self-deployable tensegrity structures for adaptive morphing of helium-filled aerostats. Archives Civil Mechan Eng 21(159). https://doi.org/10.1007/s43452-021-00292-6

Numerical Investigation of Laminar to Turbulent Boundary Layer Transition Over Airship Envelopes

Ashish Magar, Shantanu S. Gulawani, K. M. Kiran Babu, and Rajkumar S. Pant

1 Introduction

Accurate estimation of aerodynamic force coefficients is essential in the design of lighter-than-air (LTA) systems (i.e., aerostats or airships), and it significantly affects the sizing of propulsion and control systems [1, 2]. Even for shape optimization of the LTA system envelope, the estimation of aerodynamic coefficients plays a very vital role. These aerodynamic coefficients can be obtained using various techniques such as empirical formulae, experimental studies, or CFD codes. The drag coefficient for the bodies of revolutions is generally obtained using empirical formulae or co-relations based on experimental studies, such as one reported by Hoerner [3]. However, such formulations are only valid for envelope shapes with limits on a range of their fineness ratios and position of maximum diameter, as confirmed in previous studies [2, 4]. Determination of aerodynamic coefficients by experimental techniques involves a huge experimental setup and often it turns out to be economically not viable to test the full-scale model, and scaled-down models can be tested using wind tunnels. But while scaling down models, the effect of laminar to turbulent transition is compromised. To initiate the effect of laminar to turbulent transition, turbulence strips are placed on the model at an assumed point during experimental testing. This significantly affects the accuracy of aerodynamic coefficients. Wang et al. [5] have carried out the experimental testing of a scaled-down model of the *Zhiyuan-1* airship by placing the turbulence strips at transition locations obtained using empirical formulae [5].

A. Magar
SRF, CFD Virtual Reality Pvt. Ltd, Bengaluru, India

S. S. Gulawani (✉) · R. S. Pant
Indian Institute of Technology Bombay, Mumbai, India
e-mail: gulawanishantanu@gmail.com

K. M. K. Babu
Annasaheb Dange College of Engineering & Technology, Ashta, India

D. Shukla (ed.), *Lighter Than Air Systems*, Lecture Notes in Mechanical Engineering,
https://doi.org/10.1007/978-981-19-6049-9_12

Estimation of the aerodynamic coefficients using CFD techniques is possible, but the prediction of laminar to turbulent transition is still a challenge. Classical turbulence models assume flow to be fully turbulent and this affects the accuracy of aerodynamic coefficients. Solving the full 3D Navier–Stokes equation with suitable turbulence models, or direct numerical simulation (DNS) is the most accurate and comprehensive method to simulate 3D flows around an LTA vehicle. However, the DNS technique demands a huge computational resource; therefore, performing DNS is not a viable option for many real-life LTA flight envelopes. However, the Reynolds-Averaged Navier–Stokes (RANS) method of CFD can be used to accurately predict the aerodynamic coefficients by instilling the effect of laminar to turbulent transition to the CFD code. Suman et al. [6] have carried out a 3D RANS simulation of flow over the *Zhiyuan-1* airship envelope by assuming the transition point at various locations along its length and have compared the results obtained with experimental work carried out by Wang et al. [5].

The present study aims to establish the efficacy of laminar to turbulence transition over bodies of revolution using an open-source code *OpenFOAM*® for accurate estimation of the aerodynamic coefficients of envelopes of LTA systems. With this aim in mind, the study will be carried out in four steps, viz.

(1) Validation of the k-ω SST-LM turbulence model [7] with experimental results obtained by Furst et al. [8] for the flat plate cases, to establish the efficacy of the k-ω SST-LM turbulence model in the transition prediction using the *OpenFOAM*®.
(2) Establishing the efficacy of *OpenFOAM*® for carrying out CFD analysis of aerostat envelopes.
(3) Evaluation of assumed point transition analysis of *Zhiyuan-1* airship envelope and to compare the results with experimental data and the results obtained by Ref. [6].
(4) Validation of the k-ω SST-LM turbulence model for the aerodynamic analysis of the *Zhiyuan-1* airship envelope using the OpenFOAM®.

OpenFOAM® is a C++ library, originally developed by Henry Weller at Imperial College, London, in the late 1980s for solving continuum mechanics problems. *OpenFOAM*® C++ library is used primarily to create an executable, known as applications. The applications fall into two categories, viz., solvers, designed to solve a specific problem in continuum mechanics; and utilities, designed to perform tasks that involve data manipulation. One of the strengths of *OpenFOAM*® is that new solvers and utilities can be created by its users with some pre-requisite knowledge of the underlying method, physics, and programming techniques involved. *OpenFOAM*® is supplied with pre and post-processing environments. The interface to the pre- and post-processing are themselves *OpenFOAM*® utilities, thereby ensuring consistent data handling across all environments. The overall structure of *OpenFOAM*® is shown in Fig. 1.

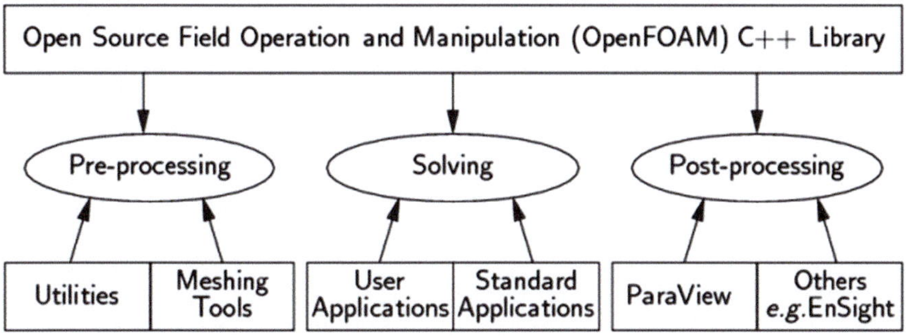

Fig. 1 Overview of OpenFOAM® structure [9]

2 Validation for Flat Plate Cases

The CFD simulations were carried out using the k-ω SST-LM turbulence model for the three cases of an experimental study carried out by European Research Community on Flow, Turbulence, and Combustion (ERCOFTAC) [8]. Figure 2 shows the geometry and the grid used for the validation study of the flat plate cases, and Table 1 lists the physical conditions specified in this study, assuming no gradient and ISA sea level conditions. The results obtained for three cases are shown in Fig. 3, which shows that the two-equation k-ω SST turbulence model is not able to capture the laminar to turbulent transition, whereas the four-equation k-ω SST-LM can capture it quite well, thus establishing the efficacy of the later model in capturing the transition between the laminar and turbulent boundary layer.

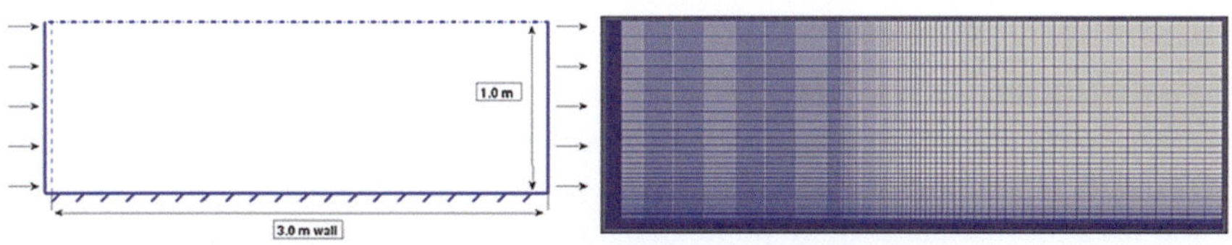

Fig. 2 ERCOFTAC flat plate test case and grid

Table 1 Three flat plate test cases

Details	T3A	T3B	T3AM
U—Inlet (m/s)	5.4	9.4	19.8
FSTI	3.3	6.5	0.87
Viscosity ratio	12.0	100	8.72

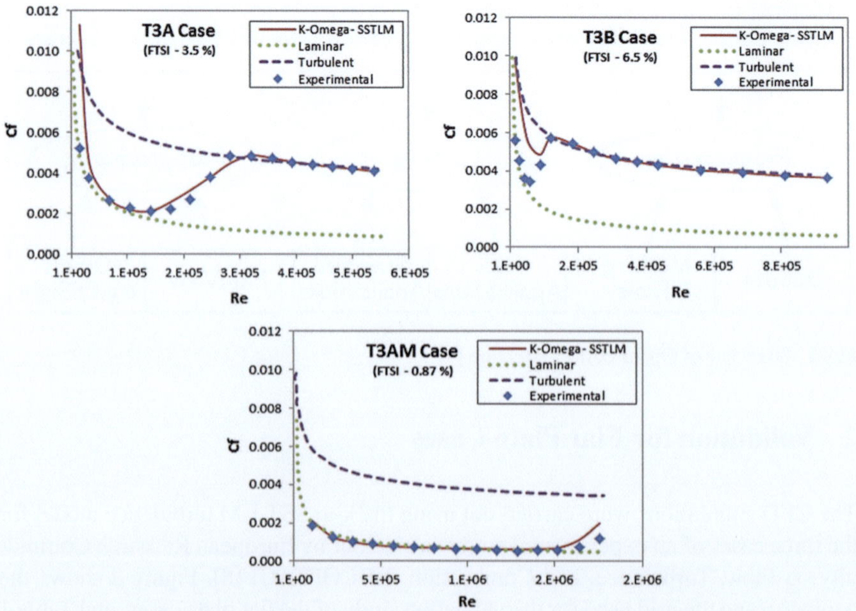

Fig. 3 Comparison of CFD results with experimental study of ERCOFTAC [8] for flat plate

3 Establishing Efficacy of OpenFOAM® for Bodies of Revolution

3.1 Drag Calculation of GNVR Envelope

Toward validation of OpenFOAM® for bodies of revolution, a GNVR envelope profile of unit maximum diameter was modeled in this study. The GNVR profile, named after its inventor the Late Prof. G. N. V. Rao of the Indian Institute of Science, is one such profile that is quite popular. This profile is a combination of three standard geometrical constructs, viz. an ellipse, an arc of a circle, and a parabola. The entire GNVR profile is parameterized in terms of maximum diameter as shown in Fig. 4. The front portion of this shape has an elliptical profile, which allows easy interfacing with the mooring system, and the rear parabolic shape ensures ease in the attachment of the fins [2]. The GNVR profile is a low-drag envelope shape for tethered aerostat deployed at low altitudes, through experimental investigations by Sundaram [10], and numerical studies using source panel methods by Narayana and Srilatha [11] and ANSYS FLUENT by Kanikdale et al. [2]. The coordinates of the profile were generated by solving the geometrical equations parametrized in terms of its maximum diameter. The length, volume, and surface area of the model were estimated to be 3.05 m, 1.43 m^3, and 7.29 m^2, respectively. The coordinates of the profile were imported into CATIA V5TM to generate a CAD model. A fluid domain

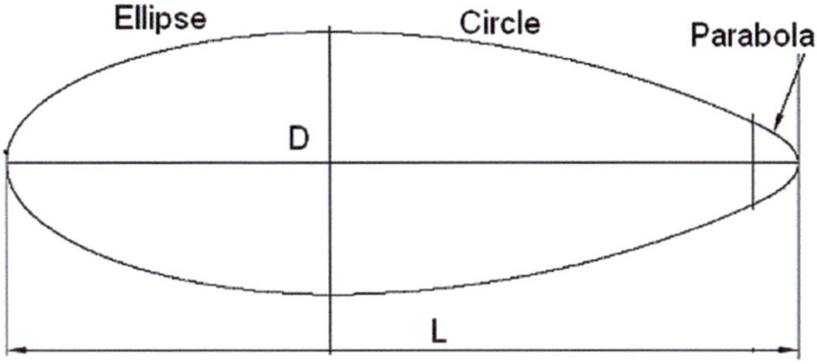

Nose Ellipse – Origin to 1.25D : $\frac{X^2}{1.25D^2}+\frac{Y^2}{0.5D^2}=1$

Middle Circle – 1.25D to 1.62D: $X^2+(Y-3.5D)^2=16D^2$

Tail Parabola – 1.62 D to 1.8D : $Y^2=1.373D(1.8D-X)$

Fig. 4 GNVR shape parameterized in terms of maximum diameter [13]

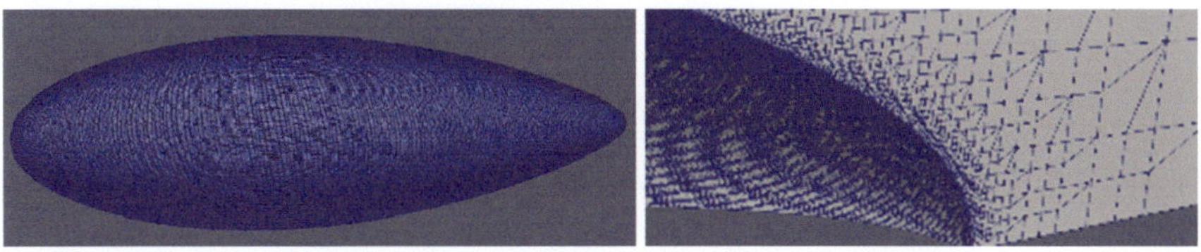

Fig. 5 Surface and boundary layer mesh over GNVR envelope

of size 35 m × 10 m × 10 m was considered sufficient for the CFD investigation. This fluid domain and the CAD model were transferred to SnappyHexMesh utility of OpenFOAM® and then mesh refinement, and snapping details are provided to generate the mesh file. Figure 5 shows the surface mesh and boundary layer mesh, respectively.

The flow physics and boundary conditions of the fluid domain used in the computational study are discussed in Table 2. In the present study, the simpleFoam solver of OpenFOAM® was used for the computational study around the envelope. The simpleFoam is a steady-state solver for incompressible, turbulent flow, using a semi-implicit method for pressure-linked equations (SIMPLE) algorithm. The standard k-ω SST turbulence model is used along with simpleFoam for the study.

Table 3 lists the parameters that were used to carry out a grid sensitivity analysis for three different grids (viz. coarse, medium, and fine).

The volumetric drag coefficient (C_{DV}) of an aerostat envelope is given by:

$$C_D=\frac{D}{\frac{1}{2}\rho U_\infty^2 V^{\frac{2}{3}}} \tag{1}$$

Table 4 compares the values of C_{DV} of the *GNVR* model for the three grid sizes.

Table 2 Flow physics and boundary conditions for numerical study of the GNVR envelope

Parameters	Value
Density	1.185 kg/m^3
Temperature	288 K
Free stream velocity	36 m/s
Reference pressure	101,325 Pa
Inlet	Velocity inlet

Table 3 Mesh information

Parameters	Coarse	Medium	Fine
Cell volumes	59,73,480	63,20,344	91,15,388
Hexahedra cells	42,92,614	46,35,278	74,12,882
Polyhedra cells	16,09,334	16,13,534	16,30,974

Table 4 Coefficient of drag data

Grid size	Volumetric drag coefficient C_{DV}
Coarse grid	0.0153
Medium grid	0.0143
Fine grid	0.0128

Figure 6 shows the pressure distribution contours; the higher pressure acting on the frontal area is clearly visible.

To establish the efficacy of *OpenFOAM*® and *SnappyHexMesh* grid generation utility for CFD investigations of GNVR envelope, the variation of pressure distribution over the envelope length was compared with that obtained in the previous studies using proprietary code (i.e., source panel method [11]) or commercial CFD

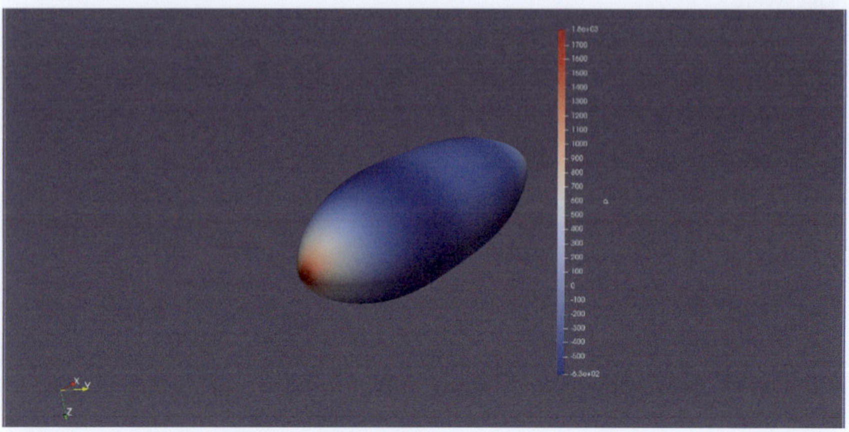

Fig. 6 Pressure contour over the GNVR envelope

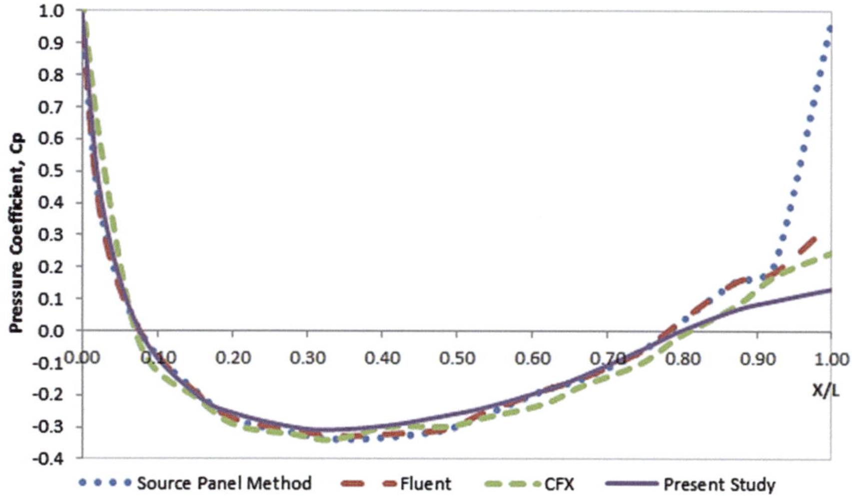

Fig. 7 Comparison of C_p distribution with literature

codes (i.e., ANSYS® *Fluent*™ [2], and *ANSYS*® *CFX*™ [12]). This comparison is plotted in Fig. 7, and a good match is seen, except in the rear end of the envelope, which is due to the ability of the code to capture the recirculation that occurs at that location due to flow separation.

3.2 Assumed Transition Point Approach for Zhiyuan–1 *Envelope*

Flow over the *Zhiyuan-1* airship envelope was modeled to evaluate the assumed transition results with the transition point located at 40%, 50%, and 60% of the length of the airship envelope. Figure 8 shows the *Zhiyuan-1* airship envelope profile, and the geometrical formulae used for deriving the profile.

Table 5 gives the results obtained for transition at 40%, 50%, 60%, and 100% of the length (i.e., fully turbulent flow); the results are compared with the results reported by Suman et al. [6].

Figure 9 shows the boundary layer mesh used for the analysis of flow over the *Zhiyuan-1* airship envelope profile.

Figure 10 plots the skin friction coefficient variation compared with the Reynolds number along the length of the *Zhiyuan-1* airship envelope.

Figure 11 shows the wall shear stress variation on the *Zhiyuan-1* airship envelope at three transition locations.

Figure 12 shows the assumed transition point comparison with the literature data from Suman et al. [6] from Fig. 12, it is clear that the coefficient of drag values is closely matching with the literature data.

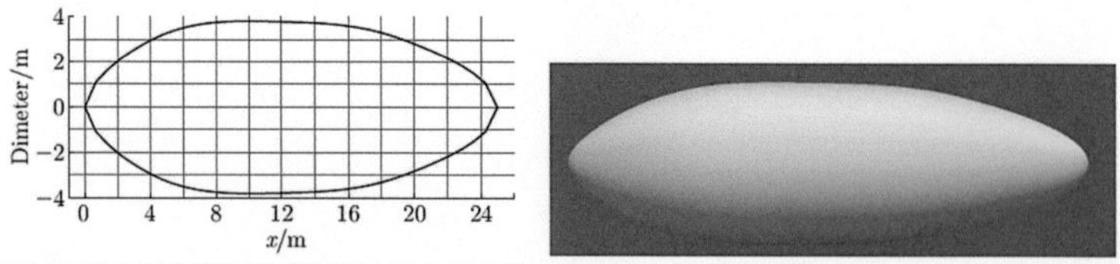

Section	Equation of parameter	Details of functions	Value
$0 \leqslant x \leqslant x_m$, $z = \frac{x}{x_m}$	$R(x) = \frac{[r_n F_1(z) + k_1 F_2(z) + G(z)]^{1/2}}{2f_r}$	$F_1(z) = -2z(z-1)^3$ $F_2(z) = -z^2(z-1)^2$ $G(z) = z^2(3z^2 - 8z + 6)$	$r_n = 0.5070992$ $k_1 = 0.2912560$ $f_r = 3.299202$
$x_m \leqslant x < x_p$, $z = \frac{(1\text{-}x)}{1 - x_m}$	$R(x) = \frac{1}{2f_r}\left[s_t^2 F_1(z) + \left(\frac{1 - x_m}{x_m}\right)^2 k_1 F_2(z) + G(z)\right]^{1/2}$	$F_1(z) = -z^2(z-1)^3$ $F_2(z) = -z^3(z-1)^2$ $G(z) = z^3(6z^2 - 15z + 10)$	$x_p = 0.7570323$ $s_t = 3.236105$ $x_m = 0.3935913$
$x_p \leqslant x \leqslant 1.0$, $z = x$	$R(x) = \frac{[c_p(1.0 - z)]^{1/2}}{2f_r}$		$c_p = 2.735107$

Fig. 8 *Zhiyuan-1* airship envelope [14]

Table 5 Comparison assumed point transition results

x/l	Parameters	Suman et al. [6]	Present study
0.4	Total C_{DV}	0.0181	0.0174
	Pressure C_{DV}	0.0133	0.0131
	Viscous C_{DV}	0.0047	0.0043
0.5	Total C_{DV}	0.015	0.016
	Pressure C_{DV}	0.004	0.005
	Viscous C_{DV}	0.011	0.011
0.6	Total C_{DV}	0.0131	0.0134
	Pressure C_{DV}	0.0041	0.0037
	Viscous C_{DV}	0.0090	0.0096
1.0	Total C_{DV}	0.025	0.025
	Pressure C_{DV}	0.006	0.006
	Viscous C_{DV}	0.019	0.020

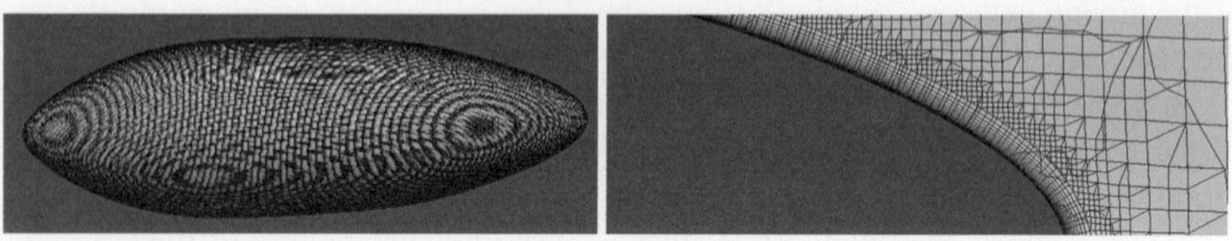

Fig. 9 Boundary layer mesh on *Zhiyuan-1* airship envelope

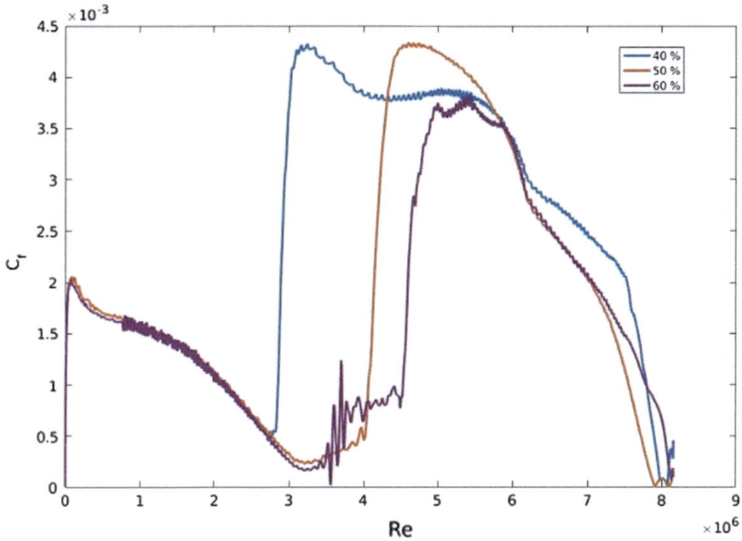

Fig. 10 Skin friction variation on *Zhiyuan-1* airship

3.3 *Evaluation of Transition Turbulence Model for* Zhiyuan-1 *Envelope*

Menter et al. [7] have proposed the k-ω SST-LM turbulence model to modify turbulent transport equations to simulate laminar, laminar to turbulent, and turbulence states in a fluid flow. This model is also sometimes known as the "$\gamma \mathrm{Re}_\theta$-SST" model, because it makes use of equations for γ and Re_θ in addition to SST's k and ω equations. The transition model consists of four equations and calculates an intermittency factor that creates or extinguishes turbulence by slowly introducing turbulent production at the laminar to turbulent transition location.

$$\frac{\partial(\rho k)}{\partial t} + \frac{\partial\left(\rho U_j k\right)}{\partial x_j} = \overline{P}_k - \overline{D}_k + \frac{\partial}{\partial x_j}\left[(\mu + \sigma_k \mu_t)\frac{\partial k}{\partial x_j}\right] \tag{2}$$

$$\frac{\partial(\rho \omega)}{\partial t} + \frac{\partial\left(\rho U_j \omega\right)}{\partial x_j} = P_\omega - D_\omega + \frac{\partial}{\partial x_j}\left[(\mu + \sigma_\omega \mu_t)\frac{\partial \omega}{\partial x_j}\right] + 2(1 - F_1)\frac{\rho \sigma_{\omega 2}}{\omega}\frac{\partial k}{\partial x_j}\frac{\partial \omega}{\partial x_j} \tag{3}$$

For intermittency, γ

$$\frac{\partial(\rho \gamma)}{\partial t} + \frac{\partial\left(\rho U_j \gamma\right)}{\partial x_j} = P_\gamma - E_\gamma + \frac{\partial}{\partial x_j}\left[\left(\mu + \frac{\mu_f}{\sigma_f}\right)\frac{\partial \gamma}{\partial x_j}\right] \tag{4}$$

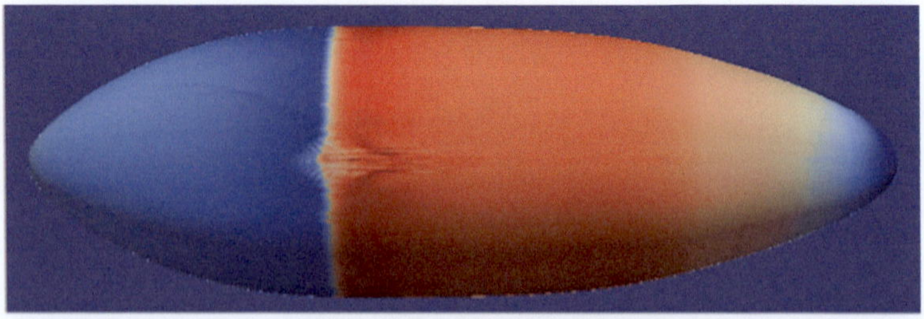

a) Transition point at 40% length

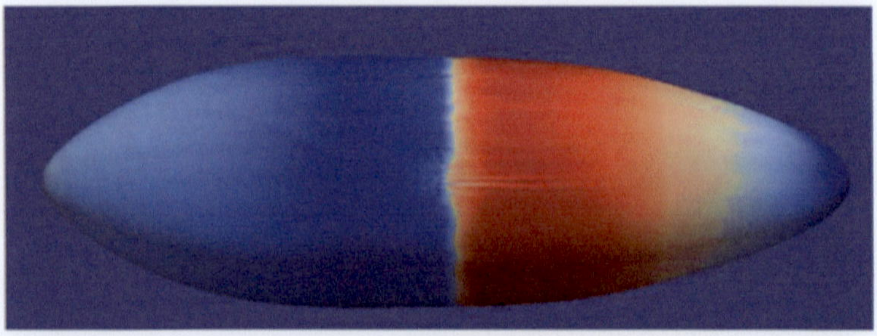

b) Transition point at 50% length

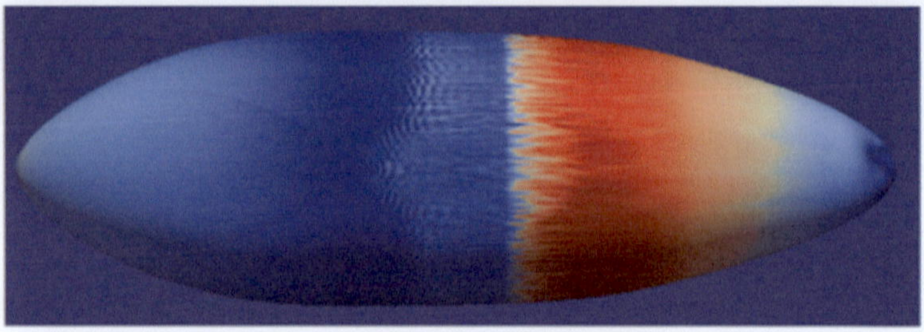

c) Transition point at 60% length

Fig. 11 Wall shear stress variation on *Zhiyuan-1* airship envelope

The intermittency equation is used to trigger the transition process; it controls the production of turbulent kinetic energy in the boundary layer.

For transition momentum thickness Reynolds number Re_θ

$$\frac{\partial\left(\rho\overline{\mathrm{Re}_{\theta t}}\right)}{\partial t}+\frac{\partial\left(\rho U_j\overline{\mathrm{Re}_{\theta t}}\right)}{\partial x_j}=P_{\theta t}+\frac{\partial}{\partial x_j}\left[\sigma_{\theta t}(\mu+\mu_t)\frac{\partial\overline{\mathrm{Re}_{\theta t}}}{\partial x_j}\right] \tag{5}$$

The transition onset Reynolds number equation avoids the additional non-local operations introduced by the experimental correlations based on free stream values, and the pressure gradient outside the boundary layer. The k-ω SST-LM model does not model the physics of the transition process. The physics of the process is contained in the experimental correlations provided to the model [7]. The key motivation of this

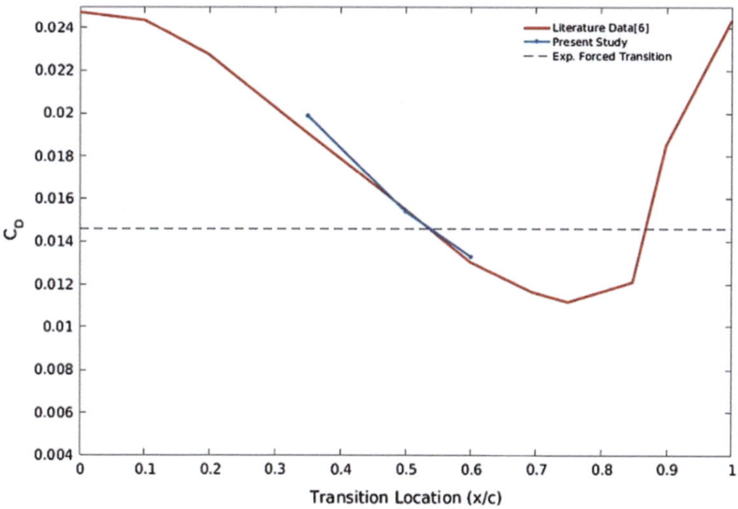

Fig. 12 Comparison of assumed transition point with Suman et al. [6]

study was to investigate the efficacy of the k-ω SST-LM transition turbulence model applied to the bodies of revolution to accurately predict the onset of the laminar to turbulent boundary layer transition and to verify the results with the assumed transition point. The reference conditions for the computations exactly match the ones employed in the experiments in [5]. The freestream velocity is 60.39 m/s, and the Reynolds number based on the volume is 2.4×10^6. Figure 13 shows the wall shear stress variation on *Zhiyuan-1* airship envelope using the k-ω SST-LM transition turbulence model.

Figure 14 shows that the variation of skin friction coefficient (C_f) along the length of the *Zhiyuan-1* airship envelope in the present study closely matches those quoted by Suman et al. [6], in which transition is assumed to occur at 52% of the envelope length.

Table 6 lists show that the values of the pressure and viscous drag coefficients obtain in this study exactly match those reported for the experimental values obtained by Wang et al. [5].

3.4 *Transition Analysis on* Zhiyuan-1 *Envelope at Angle of Attack*

Figures 15a shows the lift coefficients and drag coefficients results of the hull at a different angle of attacks (α). The transition analysis results are plotted against the free transition experimental data measured in the 3.2 m diameter wind tunnel by Wang et al. [5].

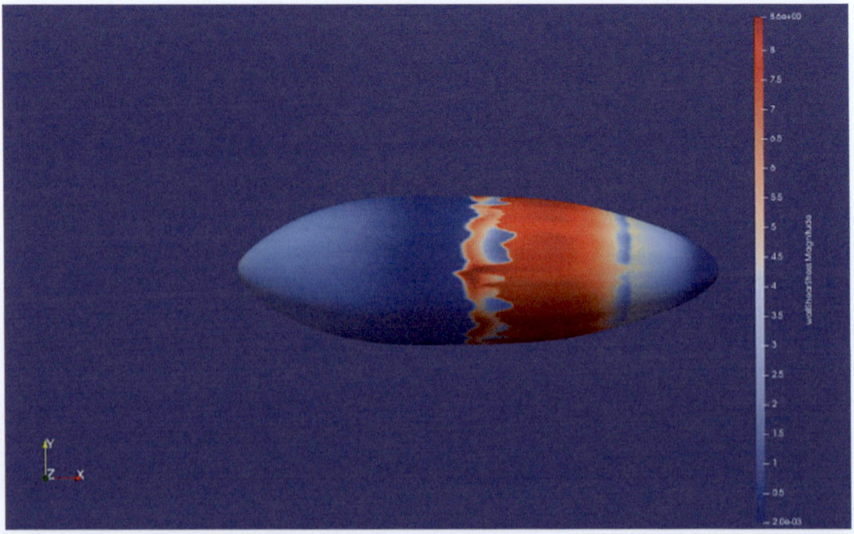

Fig. 13 Wall shear stress variation on *Zhiyuan-1* airship envelope (k-ω SST-LM model)

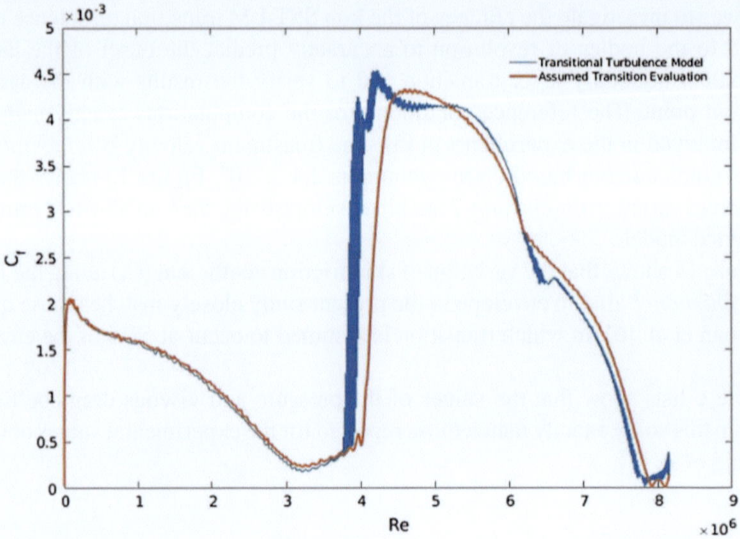

Fig. 14 Skin friction variation on *Zhiyuan-1* airship

Table 6 Comparison drag coefficient results

Parameters	Experimental data Wang et. al. [5]	k-ω SST-LM transition turbulence model
Total C_{DV}	0.015	0.015
Pressure C_{DV}	0.004	0.004
Viscous C_{DV}	0.011	0.011

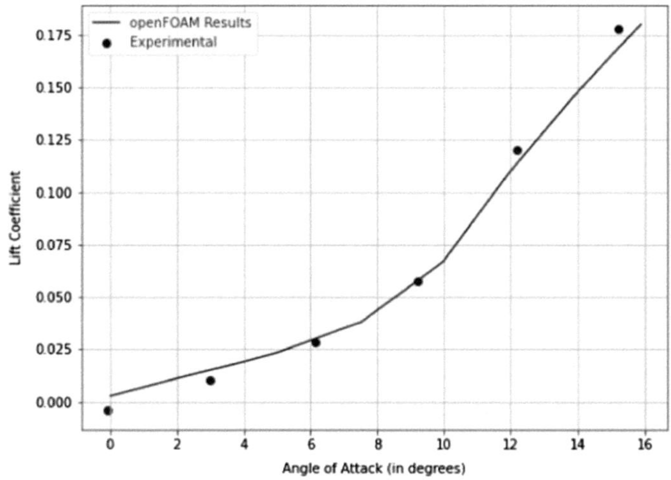

a) Lift Coefficient

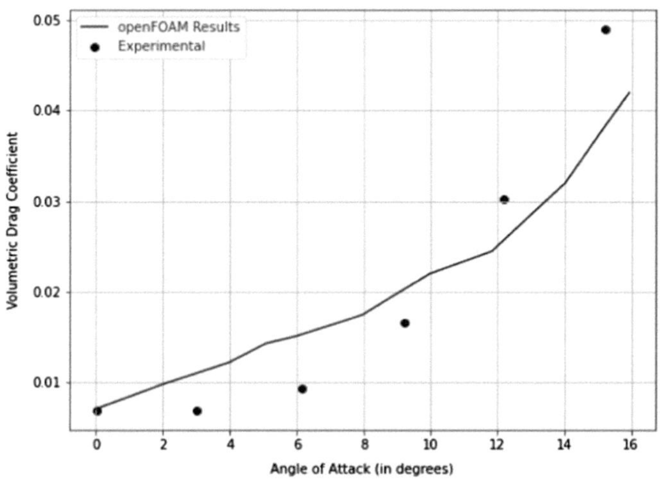

b) Volumetric Drag Coefficient

Fig. 15 Comparison of lift and drag coefficient of *Zhiyuan-1* airship hull at different angles of attack with free transition experimental data

The results in Fig. 15a show the lift coefficients of the simulation are almost the same as the experimental data. Since the lift is produced by the pressure distribution on the hull, the transition has a little effect on it. In Fig. 15b, the drag at zero angle of attack is matching with the experimental data as shown in Sect. 3.3. The variation of drag with the angle of attack, however, only matches the trend of the increasing drag coefficient. At the lower range of α, the drag coefficient increases almost linearly instead of the quadratic manner obtained in experimental results. The deviation at higher angles is suspected because of the high flow separation behind the bluff body.

4 Conclusions

The numerical investigations carried out as part of this study indicate that the two-equation k-ω SST turbulence model is not able to capture the laminar to turbulent transition, whereas the four-equation k-ω SST-LM transition turbulence model can capture it quite well, and this turbulence model can be used directly for the numerical computations to predict the onset of the laminar to turbulent boundary layer transition instead of the assumed transition approach. The study also confirms the efficacy of using OpenFOAM® V 5.0 for CFD analysis of aerostat envelopes for these operating conditions, since the pressure distribution over the envelope obtained in this study is seen to compare very well with those reported in the literature using proprietary or commercial CFD codes. For CFD analysis of airships, the analysis at a different angle of attacks is performed to study the effect of transition. The lift coefficient of the hull matches well with the experiments. The drag coefficient matches the trend of experimental data of free transition on the scaled model of the *Zhiyuan-1* airship. A further study shall include validation of transition lengths at a higher angle of attacks and efficacy of the transitional turbulence model for flows with separation.

Acknowledgements The authors acknowledge Prof. Joel Guerrero from the University of Genoa, for the learning material provided for learning OpenFOAM from the very scratch basics and would further extend a sense of gratitude toward the FOSSEE Team, IIT-Bombay, for helping us with the video tutorials. The authors also acknowledge Prof. Sawan Suman Sinha of the Department of Applied Mechanics of the Indian Institute of Technology, Delhi, for his valuable suggestions in CFD modeling.

References

1. Rajani A, Pant RS (2010) Dynamic stability analysis of a tethered aerostat. J Aircr 47(5):1531–1538
2. Kanikdale T, Marathe A, Pant RS (2004) Multi-disciplinary optimization of airship envelope shape. In: 10th AIAA/ISSMO Multidisciplinary Analysis and Optimization Conference, New York, USA
3. Hoerner SF (1965) Fluid dynamic drag, 3rd edn. Midland Park, NJ, USA
4. Schweyher H, Lutz T, Wagner S (1996) An optimization tool for axisymmetric bodies of minimum drag. In: 2nd international airship conference, Stuttgart/Friedrichshafen
5. Wang X, Fu G, Duan D, Shan X (2010) Experimental investigations on aerodynamic characteristics of the ZHIYUAN-1 airship. J Aircr 47:1463–1468
6. Suman S, Lakshmipathy S, Pant RS (2013) Evaluation of assumed-transition-point criterion in context of Reynolds-averaged simulations around lighter-than-air vehicles. J Aircr 50(2):450–456

7. Menter FR, Langtry R, Volker S (2006) Transition Modeling for general purpose CFD codes. Flow Turbul Combust 77(1):277–303
8. Furst J, Straka P, Prihoda J, Simurda D (2013) Comparison of several models of Laminar/Turbulent Transition. In: EPJ Web of Conferences, 45
9. Greenshields CJ (2022) OpenFOAM user guide version 5.0. The OpenFOAM Foundation, http://openfoam.org/, last accessed on 24th January 2022
10. Sundaram S (1998) Wind tunnel tests on 1:7 and 1:28 scale aerostat models. Project Document FE 9802, National Aerospace Laboratories, Bangalore, India
11. Narayana CL, Srilatha KR (2000) Analysis of aerostat configurations by panel methods. Project BLISS Report NAL PD 0010, National Aerospace Laboratories, Bangalore, India
12. Magar A, Kiran Babu KM, Pant RS (2017) Establishing the efficacy of Open-FOAM for carrying out CFD analysis of aerostat envelopes. In: Proceeding of National Conference on Aerial Delivery & Airborne Surveillance Systems (ADASS), ADRDE, Agra, India
13. Vijayram C, Pant RS (2010) Multidisciplinary shape optimization of aerostat envelopes. J Aircr 47(2):1073–1076
14. Liu P, Fu G-Y, Zhu LJ, Wang XL (2013) Aerodynamic characteristics of airship Zhiyuan-1. J Shanghai Jiaotong Univ (Sci) 18(6):679–687

Designing Helium-Filled Aerostats Applying Scaling Procedure, Mini-Models CANDY and Fly-Tests on SKYLAB

Jan Holnicki-Szulc, Lech Knap, Andrzej Świercz, Grzegorz Mikułowski, and Cezary Graczykowski

1 Introduction

Airships and balloons were the first men-built flying vehicles. Although their initial development has been intensive, during the last century, they have been to a large extend replaced by aircrafts. However, recent technological developments, such as production of advanced ultra-light helium-tight materials, have caused a renewed interest in aerostats. Currently, produced lightweight airships (Fig. 1a) have an ability to carry out long-term and low-energy missions at much lower costs than aircrafts [1]. They successfully fulfil various civil and military tasks, e.g. provide communication (4G/5G technology) in hardly reachable areas and serve as reconnaissance and surveillance systems or research pseudo-satellites [2]. Stratospheric airships provide local connectivity in areas of natural disasters or areas with a low level of ground infrastructure [3]. Moreover, the total costs for cargo transport or for exploration of stratosphere, mesosphere and even space observation using airships are much lower than in case of using airplanes or standard satellites [4, 5].

J. Holnicki-Szulc · A. Świercz · G. Mikułowski · C. Graczykowski (✉)
Institute of Fundamental Technological Research, Polish Academy of Sciences, Warsaw, Poland
e-mail: cgraczyk@ippt.pan.pl

J. Holnicki-Szulc
e-mail: holnicki@ippt.pan.pl

A. Świercz
e-mail: aswiercz@ippt.pan.pl

G. Mikułowski
e-mail: gmikulow@ippt.pan.pl

L. Knap
Institute of Vehicles and Construction Machinery, Warsaw University of Technology, Warsaw, Poland
e-mail: lech.knap@pw.edu.pl

D. Shukla (ed.), *Lighter Than Air Systems*, Lecture Notes in Mechanical Engineering,
https://doi.org/10.1007/978-981-19-6049-9_13

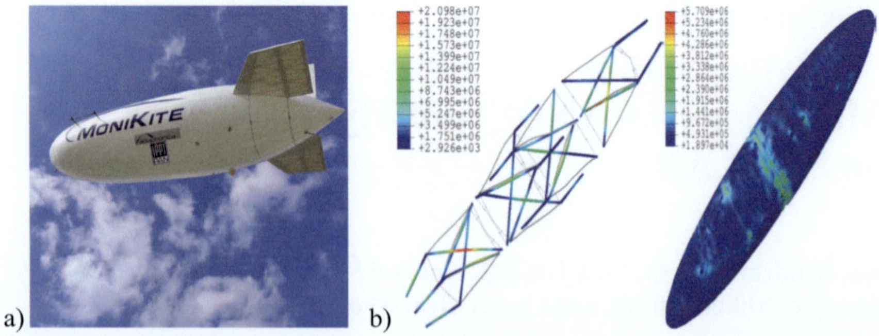

Fig. 1 **a** Classic tethered aerostat with well-designed aerodynamics developed by authors and **b** modelled aerostat-SDT with supporting tensegrity structure

Aerostat-SDT (aerostat with Self-Deployable Tensegrity) [6, 7] is an aerostat filled with gas from the storage tank with compressed helium (and storage tank with air used for ballonets) with well designed, aerodynamic shape and internal self-adaptive tensegrity-based supporting structure (Fig. 1b). Its shape can be automatically modified (increasing total volume, together with preserving aerodynamic properties and minimizing wind drift) during climbing up, from the starting level H_1 (lifted by an aircraft, rocket or by a traditional and hot-air balloon—Fig. 2), to the final altitude H_2. At altitude H_2, after monitoring mission is completed, the descent phase can be started utilizing the self-adaptive shape control of the aerostat.

The problem of optimal Aerostat-SDT design, to perform (with minimal cost of helium and energy consumption) determined mission (timing, the levels H_1 and H_2 and the payload to be lifted) with planned vertical mobility can be solved via numerical simulations, but the corresponding experimental verification can be complicated and costly. Especially, testing of V-Mobility (i.e. vertical mobility) dynamics via control of release valves for helium and ballasting air can be important.

Fig. 2 **a** Balloon SKYLAB and **b** mini-model CANDY with two ballonets

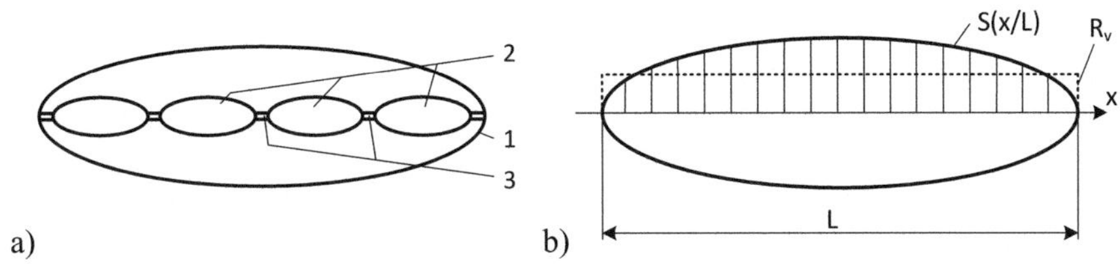

Fig. 3 **a** Mini-model aero-SDT—option with adaptive tendons: 1—external rigid shell, 2—internal softshell, 3—adaptive tendons and **b** a scheme of the scaled aerostat

2 Testing V-Mobility Via Mini-Models CANDY

V-Mobility control via gas release (ballasting air from ballonets for climbing up and helium from the other ballonet for descending) and the shape adaptation (morphing composed of the volume and aerodynamics modifications) can be tested taking advantage of a traditional, hot-air balloon (see Fig. 2a), so-called SKYLAB, towing up mini-model CANDY (see Fig. 2b), with the total mass known and also with monitored towing force F in the towing rope. Additionally, by measurements of physical parameters (temperature, external and internal pressures in the CANDY and its ballonets and mass of released gases) the buoyancy force can be determined and dynamics of vertical movement calculated.

In order to make manufacturing of mini-models easier, instead of the internal tensegrity structure, let us consider another option for controlling of the morphing process (cf. Fig. 3), so-called Aero-SDT (Self-Deployment and adaptive Tendons). The impermeable CANDY shell (2) with tight, light and very low deformable second layer (1) but equipped with adaptive strings with controllable length (3) allow effective tuning of the CANDY volume, together with the aerodynamic shape (superposed morphing).

3 Volume Measurement by Optical Technique

One of the objectives in the conducted research is to propose a procedure for determination of the physical volume of the inflated objects. Since the considered laboratory objects are fabricated in small series or even as single pieces, the repetitiveness of the production process always needs to be confirmed. Moreover, a fact is that the exact geometry of the object may rely on the changes in environmental parameters, i.e. temperature and pressure. Therefore, it becomes important to propose an adequate measuring method that is capable of identifying the modifications of the object. The measurement may be conducted by means of utilization of non-contact and optical methods, e.g. photogrammetry.

The chosen measurement method is stereo photogrammetry based on triangulation principle. Main hardware part are two cameras positioned in a precisely defined mutual position. The cameras set-up is capable to record double picture of the object that is covered with dedicated markers, in order to reconstruct the surface in 3D. Due to large scale of the measured object, a strategy of multistage photography is chosen. The multistage approach assumes recording consecutive fragments of the whole structure in order to reach the complete picture. The following step is post-processing of the recorded pictures and data analysis.

The post-processing stage consists of 3D points determination and defining the surface of the object. Within data analysis stage, a volume of the object is calculated. In the case of the ellipsoidal shape objects, the proposed calculation method may be called "truncated cone integration". The object is divided into cones with a defined spatial step, and the object's volume is calculated as a numerical integral.

The measuring task was to determine a volume of inflated structure depicted in Fig. 4a, which also presents the triangulation set-up with two cameras of 4-megapixel resolution. The object is covered with dedicated optical markers positioned on the surface in a random manner (Fig. 4b). An exemplary geometry and volume identification during the post-processing stage is depicted in Fig. 5.

Accuracy of the method is dependent on several factors that need to be taken under consideration at each stage of the process. At the stage of the picture taking, the accuracy is relevant to the number of optical markers deposited on the identified surface. At the second stage—volume calculation—the number of elementary truncated cones taken for analysis is important factor influencing the final accuracy. Both the measurements carried out, and results obtained revealed high adequacy of the proposed approach.

a) b)

Fig. 4 Experimental set-up: **a** experimental object and set-up of cameras and **b** optical markers on the surface

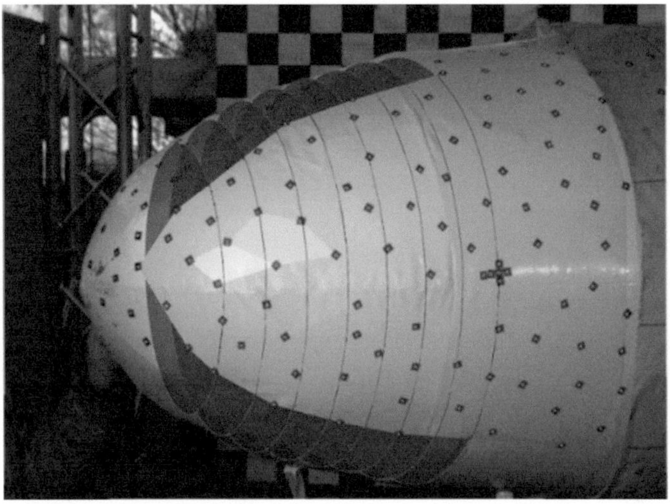

Fig. 5 Example of the left camera picture with post-processing overlayed

4 The Scaling Procedure

Assume the optimal shape of aerostat with minimized wind drift and generation of additional lifting force in windy weather (cf. function $S(x/L)$ in Fig. 3b). Let us now keep the geometrical similarity of this shape and apply the procedure of scaling up the aerostat size determined by its total length L. Then, let us define the quantity $R_v = \beta_v L$ as the radius of the roller with the same length L and the same volume of our aerostat. Analogously, let us define the quantity $R_s = \beta_s L$ as the radius of roller with the same length L and the same lateral surface of our aerostat. One can express the design volume V_0 and the design lateral surface S_0 for the configuration of the model at the starting point (at the initial level H_1, pressure difference $\Delta p = 0$ and temperature $T = T_1$) as follows:

$$V_0 = \pi(R_v)^2 L = \pi(\beta_v)^2 L^3 \tag{1}$$

$$S_0 = 2\pi R_s L = 2\pi \beta_s L^2 \tag{2}$$

By assuming the scaling process based on geometrical shape similarity rule and the scaling coefficient α as well as general dependence of aerostat volume on pressure and temperature expressed by coefficient k_v, one obtains the formula for the volume and lateral surface of the scaled aerostat:

$$V_\alpha(\Delta p, \Delta T) = \pi(\beta_v)^2(\alpha L)^3 k_v(\Delta p, \Delta T) \tag{3}$$

$$S_\alpha(\Delta p, \Delta T) = 2\pi\beta_s(\alpha L)^2 k_s(\Delta p, \Delta T) \tag{4}$$

where $k_v(\Delta p, \Delta T),\ k_s(\Delta p, \Delta T)$ are the coefficients describing the change of aerostat volume and surface, respectively, depending on gas overpressure inside the aerostat Δp and temperature change ΔT in relation to the initial temperature T_1. In the simplest case, the coefficient k_v depends linearly on gas overpressure Δp according to the formula:

$$k_v(\Delta p, \Delta T) = k_v(\Delta p) = 1 + k^{-1}\Delta p \tag{5}$$

where k is the constant determined from experimental or numerical analysis.

The total mass of the scaled aerostat can be calculated using initial lateral surface of the aerostat, its initial volume and mass of payload as:

$$M = S_\alpha(0, 0)\delta_s + V_\alpha(0, 0)(\delta_h + \delta_v) + M_p \tag{6}$$

where: δ_s—specific mass of the aerostat surface at the initial altitude H_1, δ_h, δ_v, respectively, density of helium and internal structure (per unit volume, including the mass of ballonets and air in one of them) at the initial altitude H_1 and M_p—mass of the payload to be carried up. In the consequence, the force acting on the aerostat F_α, being the difference of aerostat weight and the buoyancy force at a given altitude H, can be expressed as follows:

$$F_\alpha(\Delta p, \Delta T) = Mg(H) - V_\alpha(\Delta p, \Delta T)\delta_a(H)g(H) \tag{7}$$

where: $\delta_a(H)$—specific mass of air at the altitude H. The optimal scaling up of the CANDY aerostat in order to design Aero-SDT for the planned mission to be performed (between H_1 and H_2 and for the payload M_p) leads to determination of minimum value of scaling coefficient α for which the force acting on the aerostat is negative $F_\alpha \le 0$.

5 The Case Studies

5.1 Numerical Model

The two numerical models of the mini-aerostat (CANDY) were prepared using Abaqus software. In the first variant, a tight, non-stretchable nylon fabric partially coating a flexible aerostat membrane was applied in order to limit deformations caused by internal pressure (Fig. 6a). For comparison, the alternative model which does not contain any constraints was developed (Fig. 6b). In both cases, the aerostat models were suspended on the cable. The designed mini-aerostat is 3.06 m long and has 0.78 m in diameter. It is made of polyurethane film weighing 1.600 kg, whereas

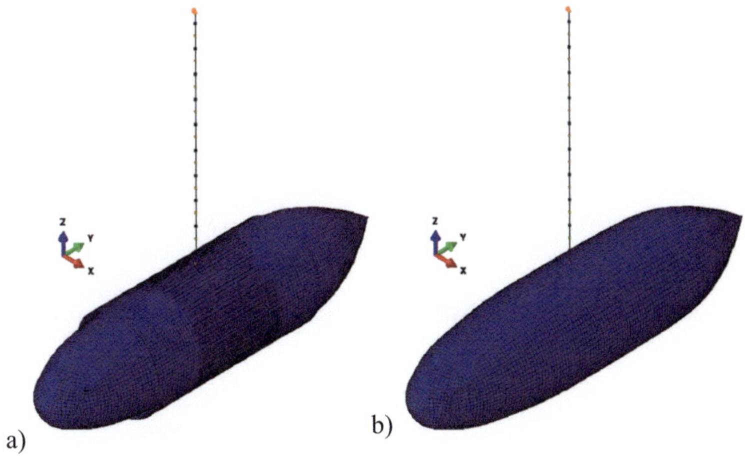

Fig. 6 Two numerical models of the suspended aerostat: **a** with constraints—non-stretchable fabric and **b** without constraints

the nylon fabric weighs 0.228 kg. The total mass of the aerostat is equal to 2.955 kg. The aerostat contains three chambers: the rear air-filled ballonet (0.098 m^3), the front air-filled ballonet (0.100 m^3) and helium-filled main chamber (0.807 m^3).

Numerical computations are based on membrane finite elements used for aerostat membrane and ballonets (11 112 elements) and for nylon fabric (5 820 elements). The rope (10 elements) is fixed at the upper end and another end is connected to the nylon fabric (Fig. 6a) or to the aerostat membrane (Fig. 6b). Changes of mass of gas in all chambers are controlled by inflators, defined for each chamber.

Independently, for comparison purposes, an experimental test was performed. In the initial configuration of the aerostat, the both ballonets were empty, whereas main chamber was filled with helium with zero overpressure. Next, some amount mass of helium was inflated into the main chamber resulting in overpressure, and then a portion of air was pumped into the front ballonet. The rear ballonet remained all the time uninflated.

The volume changes were recorded by the optic measuring technique discussed in Sect. 3. Additionally, pressure sensors were located in the front ballonet and the main chamber. The tensile force data in the suspension cable was collected by a force sensor. Obtained results of the numerical computation and the experimental test were summarized in Table 1.

The measured results given in Table 1 are coinciding with numerical simulation. In the next step, the scaling procedure presented in Sect. 4 is applied in Sect. 5.2.

Table 1 Results of numerical computation and experimental results

State	Total volume		Pressure		Force
	Measured (optic) [m^3]	Computed [m^3]	Measured [Pa]	Computed [Pa]	Measured [N]
Initial configuration	1.0102	1.0049	0.0	0.0	16.90
Main chamber inflation (helium)	1.0663	1.0548	Main chamber: 953	Main chamber: 947	16.36
Front ballonet inflation (air)	1.0697	1.0582	Main chamber: N/A	Main chamber: 1025	16.50
			Front ballonet: 830	Front ballonet: 1026	

5.2 Numerical Simulation of Aero-SDT Mission

Data obtained from experimental measurements can be used to build a numerical model of vertical motion of the CANDY demonstrator. Based on the identified model parameters and the scaling assumptions presented in the previous sections, sample analyses of the operational capabilities of the scaled aerostats were performed. Selected results are summarized in Table 2. During the simulation, a mission was assumed to lift a 0.5 kg payload to the maximum height that can be achieved with a given type of scaled structure.

The presented results show that both the initial structure and the structure with lengths scaled by the $\alpha = 1.7$ factor do not have the ability to ascend at the assumed density parameters of the materials used. Only the scaled structures with α value greater than 1.7 allow the aerostat to realize the mission.

Figure 7 shows the results of the aerostat mission for the scale factor of $\alpha = 1.8$. The left chart presents the exemplary altitude changes in time. One can see the maximum operating altitude which is just over 500 m. On the other hand, in Fig. 7b, the variation of the buoyancy force and the total weight as a function of time are shown. During the initial phase of movement, it is visible that the buoyancy force is greater than the weight causing the aerostat to ascend to its maximum height. This operation is accompanied by drop of the buoyancy force. Despite the decrease in atmospheric pressure, the coating material above a certain volume must be stretched which limits the change in volume. The aerostat maintains altitude when it is in equilibrium. Then, after releasing a small amount of helium, the buoyancy force is smaller than the weight, which causes the aerostat to descend. It is worth noting that as the altitude decreases, the atmospheric pressure increases which additionally decreases the volume and buoyancy force.

Table 2 Results of numerical simulations of aero-SDT mission

α	V/V_0	L	D	V	A	Specific mass δ_v	Specific mass δ_s	k	Payload	Buoyancy force (initial)	Weight	Maximal altitude
1.0	1.0	3.06	0.79	1.01	6.14	0.30	0.20	18,868	0.50	12.45	21.70	0.00
1.7	4.9	5.20	1.35	4.94	17.76	0.30	0.20	6739	0.50	60.90	62.75	0.00
1.8	5.8	9.36	2.42	5.86	19.91	0.30	0.20	5896	0.50	72.35	71.27	520.00
2.0	8.0	6.12	1.58	8.04	24.58	0.30	0.20	3494	0.50	99.30	90.55	1760.00

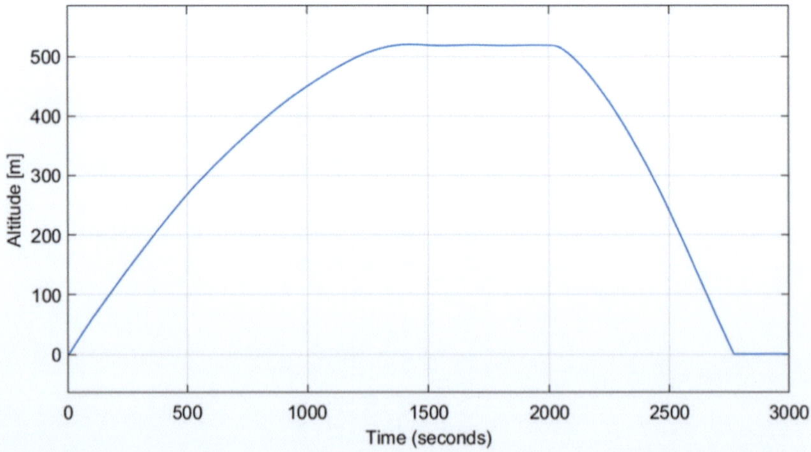

a)

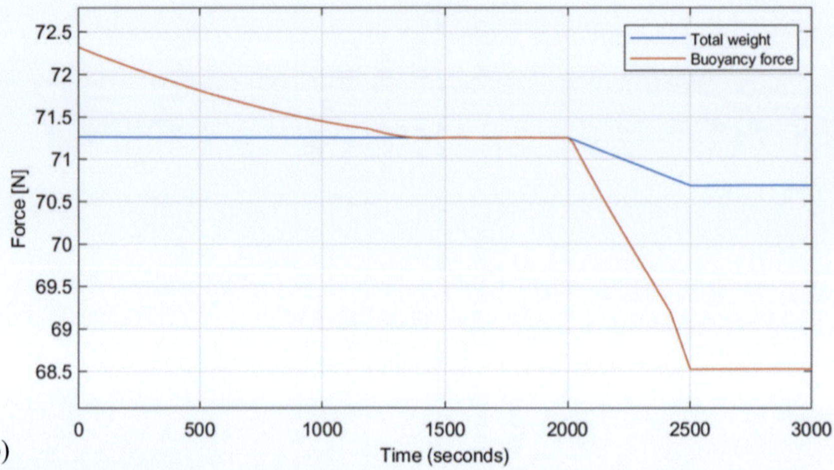

b)

Fig. 7 Simulation results of the mission for the aerostat scaled by $\alpha = 1.8$: **a** altitude changes and **b** changes in buoyancy force and aerostat weight

6 Discussion and Conclusions

The important limitations of aerostats design (especially their ability to carry required payloads or to achieve assumed altitudes) are the problems related to selection of appropriate aerostat internal structure and volume affecting the necessary lift force. A significant change in payload or target altitude causes the requirement of resizing the aerostat. Although it is possible to use oversized aerostats, operating them is associated with higher costs (e.g. the need to use much larger mass of expensive gases or increased fuel consumption). For this reason, we address the novel concept of designing the customized aerostat based on a mini-models (CANDY) which can be more easily verified and tested. If such a design is properly functioning, the dedicated scaling procedure for aerostats dimensions can be applied. We also present preliminary results of numerical tests of a proposed design. The final validation of the proposed procedure of designing helium-filled aerostats and mini-models CANDY and the scaling procedure will be experimental fly testing in near real conditions using SKYLAB.

Acknowledgements The authors acknowledge the support of the National Centre for Research and Development and the National Science Centre, Poland, granted in the framework of the TANGO 4 programme (project TANGO-IV-C/0001/2019-00).

References

1. Recent Development Efforts for Military Airships, The Congress of the United States, Congressional Budget Office (CBO), November 2011, https://fas.org/irp/program/collect/cbo-airship.pdf
2. Saleh S, Weiliang HE (2018) New design simulation for a high-altitude dual-balloon system to extend lifetime and improve floating performance. Chin J Aeronaut 31(5):1109–1111
3. Loon.com. The Stratosphere: High Altitude, Higher Ambitions, https://loon.com/resources/content-library. Accessed on June 21 2021
4. Ghanmi A, Sokri A (2010) Airships for military logistics heavy lift. Defence R&D Canada, Centre for Operational Research and Analysis, DRDC CORA TM 2010-011, 2010, https://cradpdf.drdc-rddc.gc.ca/PDFS/unc92/p532881.pdf
5. Future Aerostat and Airship Investment Decisions Drive Oversight and Coordination Needs (2012) Report to the Subcommittee on Emerging Threats and Capabilities, Committee on Armed Services, U.S. Senate, GAO
6. Holnicki-Szulc J, Świercz A, Kostro S, Knap L, Graczykowski C (2021) A concept of the SDT (self-deployable tensegrity) structure for the rapid and precise lifting of helium aerostats, especially into the stratosphere, Patent No. 3770352, EPO
7. Knap L, Świercz A, Graczykowski C, Holnicki-Szulc J (2021) Self-deployable tensegrity structures for adaptive morphing of helium-filled aerostats. Arch Civ Mech Eng 21(159):1–18. ISSN: 1644-9665. https://doi.org/10.1007/s43452-021-00292-6

Effect of Reynolds Number on the Aerodynamic Characteristics of Leading-Edge Protuberanced Airship Fin

S. Arunvinthan, C. Hari Babu, V. Manoj, and S. Nadaraja Pillai

1 Introduction

Lighter than aircraft (LTA) popularly known as airship is one of the first aircraft that realized humans dream of flight. Jean-Baptiste Meusnier proposed the first technical design of an airship featuring ellipsoidal shape with control surfaces like elevator and rudder in 1784 but lightweight powerful engine hindered his flight. Later with the advent of steam engine technology, during the year 1852, Henri Giffard incorporated steam engine technology in to airships for the first time and flew around 17 miles [1]. Even though these are some of the pioneering milestones in the history of airships, the golden age of airships began in 1900 with the launch of LZ1 Luftschiff Zeppelin of the German army. Subsequently, British and US forces began imitating its design and launch look alike airships named R-33, R-34 and Shenandoah (ZR-1) in the first half of 1900s. But with the advancements in the field of fixed wing aircraft progressing along with the occurrence of airship accidents including 1937 Hindenburg aircraft fire led to the downfall of the airships. Although the use of airships has been in recession for the past several years with the possibility of applying airships for military high-altitude long endurance surveillance and stationary airships design proposal as standby satellites has spurred interest among the researchers and this indicates the resurgence of the airship. Studies suggest that airships have wide variety of applications in telecommunication sector [2], surveillance (Stratospheric observation [3]; environmental monitoring [4]), terrain mapping [5] etc. Researchers realized

S. Arunvinthan (✉) · S. N. Pillai
Turbulence and Flow Control Lab, School of Mechanical Engineering, SASTRA Deemed to Be University, Thanjavur, Tamil Nadu 613401, India
e-mail: arunvinthan@mech.sastra.edu

C. H. Babu · V. Manoj
Department of Aeronautical Engineering—SMACE, Kalasalingam Academy of Research and Education, Krishnankoil, Tamil Nadu 626126, India

D. Shukla (ed.), *Lighter Than Air Systems*, Lecture Notes in Mechanical Engineering,
https://doi.org/10.1007/978-981-19-6049-9_14

that the flight behaviour of the airship depends upon the interaction between aerostatics and aerodynamics. Aerodynamics of the airship is one of the fundamental issues which have been investigated for many years but it has not been perfected yet. Airship designers have utilized wind-tunnel data to understand airship dynamics to a greater extent as it facilitates better design for airships. Such wind-tunnel test over the airships performed by Freeman [6] revealed that the tail fins not only produce lift force but also helps stabilizes the airship. Freeman identified that the hull of an airship experiences a nose-up pitch moment at nonzero angles of attack because of the normal force distribution at the front and the rear. Further, Freeman [7] reported that the fins attached to the hull of the airship produces 30–40% of the overall lift of an airship. Based on the wind-tunnel test performed by Curtiss [8], next to hull the second primary source of drag is due to the fins and the fins contribute around 7–27% of the total drag of the airship. Li et al. [9] confirmed this statement. Even though several studies were conducted on the drag reduction of the hull of an airship, to the extent of authors knowledge very few studies have been reported on the drag reduction of the fins of an airship. As we know that, nature offers the best solution to real-life problems as it has been optimized by evolution, it's time to look back to nature for answers. It is not new that aerodynamic designers draw inspiration from nature. In this present study, the nature-based inspiration has been drawn from humpback whale flippers. Frank E. Fish [10, 11] initiated the research on the flippers of the humpback whales and found out that the unique leading-edge protuberances present over its flippers offer aerodynamic benefits. To identify the uniqueness of such flippers with leading-edge protuberances, Fish conducted study over idealized flipper models with and without LEP and reported that flipper models with LEP experiences 25% more airflow than the flipper model without LEP. Subsequently, Fish [12–15] published several papers to his credit and also patented this technology. Further, wind-tunnel studies carried out by Miklosovic et al. [16] confirmed this behaviour. Following which, several researchers like Johari et al. [17], Custodio et al. [18] and Zhang et al. [19] conducted studies on leading-edge protuberances over NACA 63(4)-021 which closely resembles the aerofoil flippers and reported that the leading-edge protuberanced model outperforms the conventional straight blade model and is effective in drag reduction. Even though several studies have been performed over the LEP models, it should, however, be noted that no explicit study has been performed to investigate the effect of Reynolds number on the aerodynamic characteristics of such model to the extent of authors knowledge. Additionally, this will be the first of its kind in utilizing such LEP fins for airships to augment its aerodynamic efficiency by drag reduction. Recent study by Arunvinthan et al. [20] also confirmed that the modified LEP model also performs efficiently at various turbulence intensities. It is speculated that utilizing such modified LEP fins not only provide better aerodynamic performance characteristics but also aid in stabilized airship at various turbulence intensities.

2 Computational Methodology

NACA 63(4)-021 aerofoil profile has been chosen as the test model to evaluate the aerodynamic performance characteristics of leading-edge protuberanced (LEP)-based fin section for LTA. The NACA 63(4)-021 aerofoil has been chosen based on the framework of the previous researchers since it closely resembles the flippers of the humpback whale which acted as a source of bio-inspiration for the same. The leading-edge protuberances were modelled based on the sinusoidal pattern as outlined by Arunvinthan et al. [20] in his previous study. A schematic representation of the LEP test model is shown in Fig. 1. In this study, the primary focus is to identify the influence of the effect of Reynolds number on the aerodynamic characteristics on the leading-edge protuberanced fin section alone and hence, the variation of amplitude and wavelength of the leading-edge protuberances is kept constant throughout the study. Modified test model with leading-edge protuberances featuring amplitude of 0.12 °C and wavelength of 0.75°C is considered in this present study. To identify the Reynolds number effects, the test models were tested at various angles of attack ranging from 0° to 45° in increment of 5° at various Reynolds number ranging between 10^2 and 10^6.

All the test models considered in this study were modelled using GAMBIT and analysed using commercial flow solver-ANSYS FLUENT. Based on the framework of the previous researchers, it has been identified that the first and the second order upwind methods were employed to solve the governing equations using semi-implicit method for pressure linked equation (SIMPLE). Standard $K - \epsilon$ turbulence model has been used for the numerical investigation as its two-equation model exhibits excellent predictive capability for problems dealing with aerofoil surfaces with least computational effort and challenges. In this study, the flow convergence criteria have

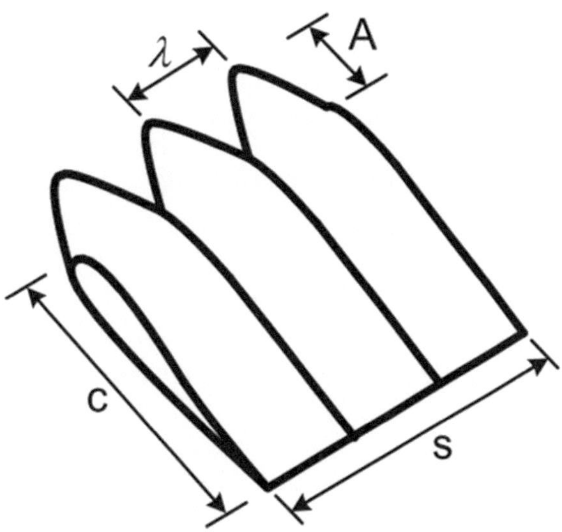

Fig. 1 Schematic representation of the modified LEP test model

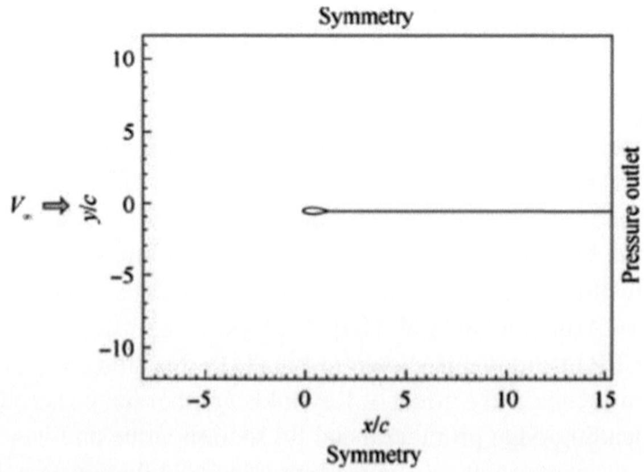

Fig. 2 Schematic representation of computational domain and boundary conditions

been set as 1×10^{-10} for all the computed quantities. The coordinate system is shown in the figure with the origin at the leading-edge point of the test model and correspondingly X, Y and Z axis measure the streamwise, lateral and spanwise directions, respectively. The fin model considered in this study possess a mean chord length 1 m. The length of the computational domain along the chordwise distance (L_C), normal to the chord (L_N) and the length of the computational domain in the lateral direction (L_Z) is of size 22.5C×20C×10C. The computational domain is designed large enough to rule out any unnecessary disturbances created by the boundaries. As this study involves complicated flow patterns over the modified LEP test model, it necessitated high-quality grids near the body surface and hence to generate an accurate mesh, algebraic initial point distribution has been used for surface grids. Likewise, grid independence test has been carried out for multiple sets of grids (0–1 million elements) with increasing mesh density of 0.2 million. Finally, 0.8 million elements were chosen as increasing the mesh density beyond this point exhibits independence in the results. In this study, the uniform inlet velocity boundary condition has been assumed on the inlet and the outlet is defined as the pressure outlet. No-slip boundary condition is set for the test model with LEP. Symmetry condition is set for lateral directions, respectively. A schematic representation of computational domain and boundary conditions is shown in Fig. 2.

3 Results and Discussion

3.1 Effect of Re on the Lift Characteristics Curve of LEP Fin

Figure 3 shows the variation of coefficient of lift (C_L) versus angle of attack (α) for the leading-edge protuberanced (LEP) test model at different Reynolds number. The results provide a quantitative measure of the aerodynamic performance of the LEP models subjected to Reynolds number ranging between 10^2 and 10^6 at various angles of attack ranging from 0° to 45°. The effect of Re is evident from the lift coefficient plots that the characteristics regimes of flow over the leading-edge protuberanced test model shifts appreciably at different Reynolds number. It is evident from the figure that Re $= 10^3$ exhibits the maximum lift coefficient in comparison against the other test Reynolds number. From the lift coefficient curve, it could be reported that the modified LEP test model exhibits better performance at low Reynolds number. At Re $= 10^3$, it could be seen that the coefficient of lift (C_L) increases linearly with the increase in the angle of attack till $\alpha = 20°$ with a lift curve slope of 0.10 deg^{-1}, which then reduces to 0.6 deg^{-1} between 25° and 30°. Beyond which the increase in the coefficient of lift (C_L) is slowed down to 0.03 deg^{-1} between $30° \leq \alpha \leq 35°$, following which the stall phenomenon occurs with a slight dip in the lift curve. The difference in the lift curve slope experienced by the same LEP model subjected to different Reynolds number clearly indicates a change in the qualitative flow structure caused by the variation in Re. For instance, it can be seen from Fig. 3 that the same LEP model when subjected to Re $= 10^6$ exhibits a lift curve slope of 0.10 deg^{-1} between $10° \leq \alpha \leq 15°$, following which the lift curve slope reduces to 0.02 deg^{-1} between $20° \leq \alpha \leq 25°$. It is of interest to note that the qualitative flow structure occurring over the same LEP model subjected to different Re tends to reduce the maximum lift coefficient. For instance, the maximum lift coefficient observed for the LEP test model when subjected to Re $= 10^3$ is 3.44 at $\alpha = 40°$, whereas at Re $= 10^6$ the maximum coefficient of lift reduces to 2.11 at $\alpha = 30°$. Henceforth, it is worth noting that with the increase in the Reynolds number, not only the decrease in the maximum lift coefficient is observed, but the angle of attack at which the maximum lift coefficient is attained is also significantly reduced. Even though, based on the lift coefficient plot, it can be claimed that the modified LEP test model is good at low Reynolds number compared to high Reynolds number, it should, however, be noted that at high Re, the modified LEP model offers sustained lift coefficient over large range of angle of attack. For instance, at Re $= 10^6$ the test model offers sustained lift coefficient between $20° \leq \alpha \leq 45°$ which is way higher than the conventional straight fin model for the lighter than air aircraft. To better understand the aerodynamic characteristics of this modified LEP test model, it becomes quintessential to understand the drag and the aerodynamic efficiency as well.

3.2 Effect of Re on the Drag Characteristics Curve of LEP Fin

Figure 4 represents the variation of coefficient of drag at every test angle of attack. Contrary to the lift coefficient plots the modified LEP test model subjected to high Reynolds number exhibits the lowest drag coefficient in comparison against the low Reynolds number. Therefore, it can be reported that the increase in the Reynolds number tends to decrease the drag coefficient between 10^2 and 10^6 at various angles

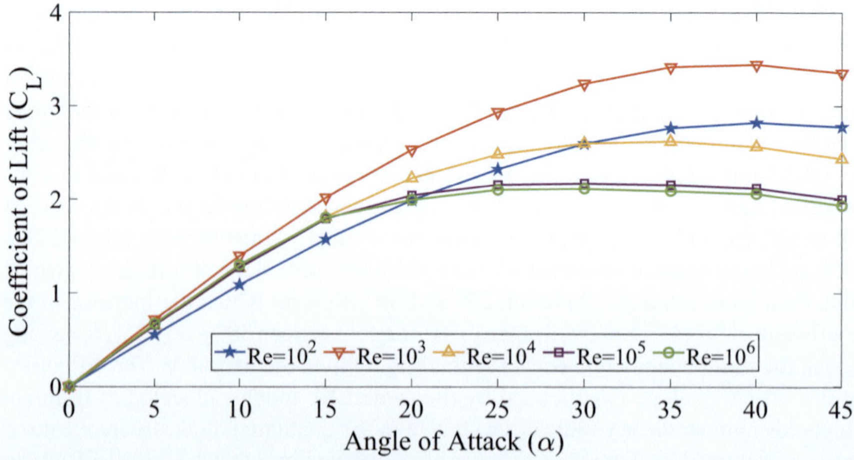

Fig. 3 Coefficient of lift (C_L) versus angle of attack (α) for LEP fin

of attack. It is believed that the increase in the momentum provided by the increase in the flow velocity accompanied with the increase in the Reynolds number might be the plausible reason behind the decrease in the drag coefficient with the increase in the Re. One might speculate that the decrease in the drag coefficient can be caused by the delayed flow separation caused by the increase in Reynolds number.

To get further insight in to this, the surface pressure distribution over the modified leading-edge protuberanced models needs to be investigated. Since the modified leading-edge protuberanced model is not a constant chord model, the surface pressure distribution needs to be investigated at various chordwise locations namely the peak

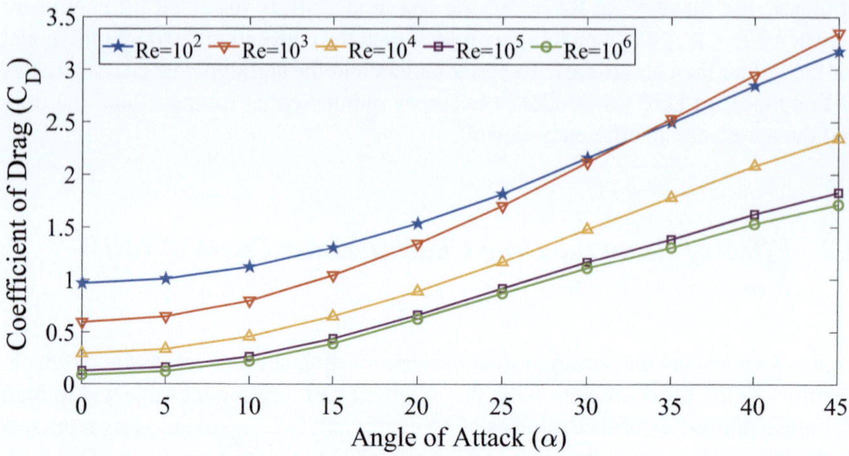

Fig. 4 Coefficient of drag (C_D) versus angle of attack (α) for LEP fin

where the chord maxima occur, mid and trough where the chord minima occur. As stated above, the surface pressure distribution over the modified leading-edge protuberanced models was obtained at peak, mid and trough sections and are displayed in Fig. 5a–c, respectively, for $\alpha = 10°$. In order to understand the influence of the Reynolds number effect on the aerodynamic characteristics, the peak section of the test model subjected to different Reynolds number is plotted in same graph. Likewise, the mid and trough section of the test model subjected to different Re is also plotted in individual graphs for better understanding. In addition, this will help us understand the flow physics to a greater extent. It is evident from the surface pressure distribution of peak, mid and the trough section, that the trough exhibits the maximum negative suction pressure signifying that the maximum velocity is attained over the trough section compared to the peak and the mid. Since the primary aim of the study is to identify the effect of Re, the variation of surface pressure distribution with Re is then observed. Generally, the oncoming flow is bifurcated at the leading edge of an aerofoil and then one flow will move over the upper surface of an aerofoil and the other one will move over the lower surface of an aerofoil. The flow which is moving over an upper surface of the aerofoil gradually accelerates indicated by the negative suction pressure (i.e. the favourable pressure gradient) and then, the pressure keeps on increasing with the increase in the x/C till the trailing edge and achieves pressure recovery (i.e. tends to get equal to the ambient pressure zero). It is identified from the graphs that the Re influences the flow characteristics prevailing over the upper surface of the aerofoil. In other words, with the increase in Re, the chordwise distance taken to achieve the maximum negative suction pressure (favourable pressure gradient) reduces. For instance, it can be seen from the Fig. 5b that at surface distribution of mid sections, at Re $= 10^2$ the maximum negative suction peak is reached at $x/C = 0.31$ whereas for Re $= 10^3$ it happens at $x/C = 0.24$ and for Re $= 10^4$ the peak suction pressure is at $x/C = 0.20$, respectively. It is of interest to note that at Re $= 10^5$ and 10^6 there is no advancement in the chordwise distance rather the increase in the maximum negative suction pressure is observed. At both Re $= 10^5$ and 10^6, the peak negative suction pressure is observed at $x/C = 0.13$. Similar trend line has been observed in the peak and the trough sections also. Therefore, it becomes clear that with the increase in the Re, the maximum negative suction pressure (favourable pressure gradient) tends to move towards the vicinity of the leading edge, thus providing reduced drag coefficient. As aerodynamic efficiency not only depends on drag reduction, the ratio of lift to drag (i.e. aerodynamic efficiency) needs to be understood for better insight.

The aerodynamic efficiency (L/D) vs. angle of attack is presented in Fig. 6. It is evident from the figure that Re $= 10^6$ exhibits peak aerodynamic efficiency over the test angles of attack especially in the pre-stall region. Following which Re $= 10^5$ exhibits the second peak and successively drops down to Re $= 10^2$ finally. This clearly shows that the modified LEP test model exhibits good aerodynamic efficiency at Re $= 10^5$ and 10^6 and hence could be used for LTA fin sections. The surface pressure distribution plots are further investigated in detail to observe more peculiar phenomenon like laminar separation bubble, etc., and the presence of the same if exists will be reported in the conference.

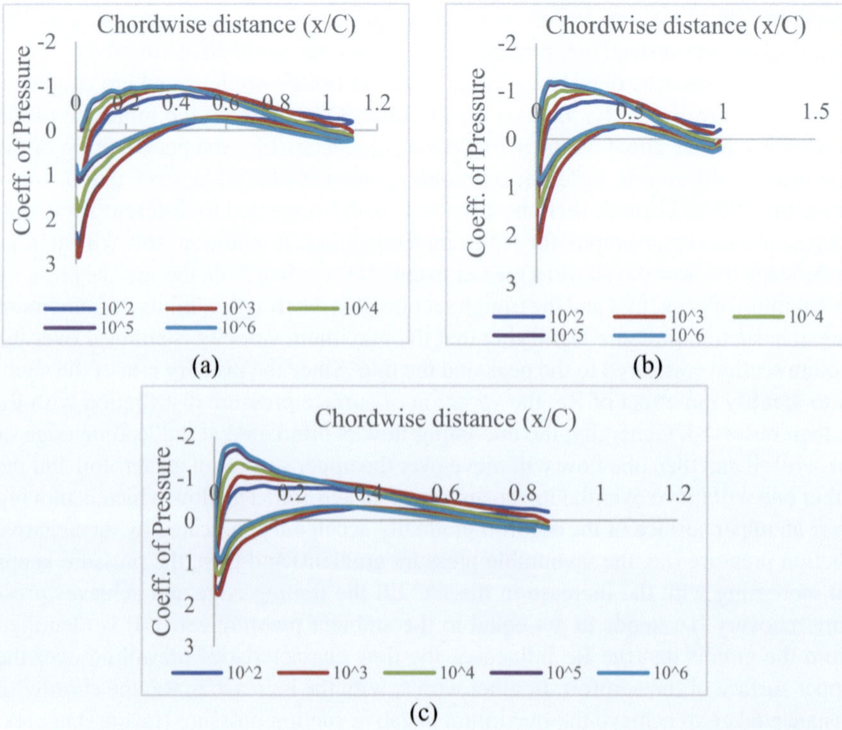

Fig. 5 Coefficient of pressure (C_p) versus chordwise distance (x/C) for **a** peak, **b** mid and **c** trough

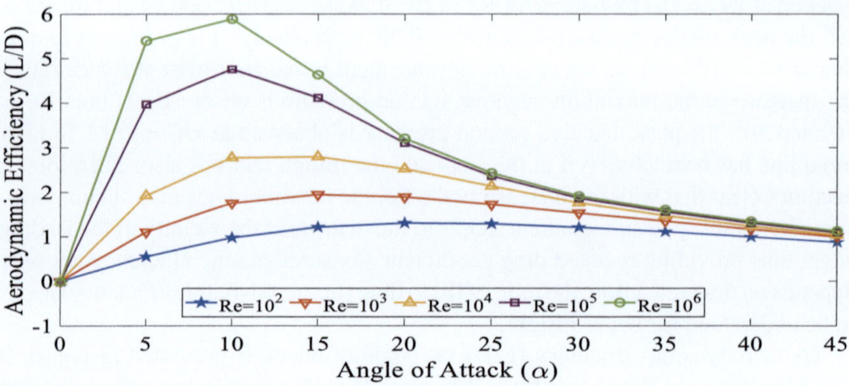

Fig. 6 Aerodynamic efficiency (L/D) versus angle of attack (α)

4 Conclusion

In this paper, the computational investigation of the effect of Reynolds number (Re) on the aerodynamic characteristics of the leading-edge protuberanced (LEP) fins at wide range of angles of attack ranging from 0 to 45° is computationally evaluated. The aerodynamic force coefficients, aerodynamic efficiency and surface pressure distribution were investigated in detail and based on the results the following conclusions were made:

1. Aerodynamic force coefficients like (C_L) and (C_D) of the leading-edge protuberanced fin section are significantly influenced by the Reynolds number.
2. Leading-edge protuberanced fin section subjected to $Re = 10^6$ shows the minimum drag coefficient and maximum aerodynamic efficiency when compared against the other test cases.
3. With the increase in the Reynolds number, the favourable pressure gradient (i.e. peak negative suction pressure) tends to move towards the vicinity of the leading edge, thus providing better aerodynamic characteristics and reduced drag.
4. Surface pressure distribution over the peak, mid and the trough section shows that the peak negative suction pressure exists at trough region, thus showing that majority of the flow goes through the trough region, thus inducing a spanwise pressure gradient among the upper surface of the aerofoil itself.

To gain better understanding, the study has to be made at multiple velocities in the same Reynolds number range itself. Attempts can be made in future to experimentally evaluate the aerodynamic characteristics of such LEP fins and flow visualization studies could also be performed to further ascertain this underlying flow physics.

References

1. Gerken LC (1990) Airships: history and technology. Chula Vista, CA: American Scientific Corporation; pp 1–19. Wilson JR (1990) A New era for airships. Aeros
2. Lee Y-G, Kim D-M, Yeom C-H (2006) Development of Korean high altitude platform systems. Int J Wireless Inf Netw 13(1):31–42
3. Lee S, Bang H (2007) Three-dimensional ascent trajectory optimization for stratospheric airship platforms in the jet stream. J Guid Control Dyn 30(5):1341–1352
4. Dorrington GE (2005) Development of an airship for tropical rain forest canopy exploration. Aeronaut J 109:361–372
5. Hygounenc E, Jung I, Soueres P, Lacroix S (2004) The autonomous blimp project of LAAS-CNRS: achievements in flight control and terrain mapping. Int J Robot Res 23(4–5):473–511
6. Freeman HB (1933) Force measurements on a 1/40-scale model of the U.S. Airship Akron. NACA TR-432
7. Freeman HB (1934) Pressure distribution measurements on the hull and fins of a 1/40-scale model of the U.S. Airship Akron. NACA TR-443
8. Curtiss HC, Hazen DC, Putman WF (2003) Experimental investigations on hull–fin interferences of the LOTTE airship. Aerosp Sci Technol 7(8):603–610
9. Li Y, Nahon M, Sharf I (2011) Airship dynamics modelling: a literature review. Prog Aerosp Sci 47:217–239

10. Fish FE (1994) Influence of hydrodynamic-design and propulsive mode on mam-malian swimming energetics. Aust J Zool 42:79–101
11. Fish FE, Battle JM (1995) Hydrodynamic design of the humpback whale flipper. J Morphol 225:51–60
12. Fish FE, Lauder G (2006) Passive and active flow control by swimming fishes and mammals. Annu Rev Fluid Mech 38:193–224
13. Fish FE, Howle LE, Murray MM (2008) Hydrodyn flow control mammals. Integr Comp Biol 48:788–800
14. Fish FE, Weber PW, Murray MM, Howle LE (2011) Marine applications of the biomimetic Humpback whale flipper. Mar Technol Soc J 45:198–207
15. Fish FE, Weber PW, Murray MM, Howle LE (2011) The tubercles on humpback whales flippers: application of bio-inspired technology. Integr Comp Biol 51:203–213
16. Miklosovic D, Murray M, Howle LE, Fish FE (2004) Leading-edge tubercles delay stall on humpback whale (megaptera novaeangliae) flippers. Phys Fluids 16:L39–L42
17. Johari H, Henoch CW, Custodio D, Levshin A (2007) Effects of LEP on airfoil perf. AIAA J 45:2634–2642
18. Custodio D, Henoch C, Johari H (2012) Aerodynamic characteristics of finite-span wings with leading-edge protuberances. In: Proceedings of 50th AIAA aerospace sciences meeting including the new horizons forum and aerospace exposition, Nashville, Tennessee. AIAA. 54, pp 1–12
19. Zhang M, Wang G, Xu J (2013) Aerodynamic control of low-Reynolds-number airfoil with leading-edge protuberances. AIAA J 51:1960–1971
20. Arunvinthan S, Nadaraja Pillai S, Cao S (2020) Aerodynamic characteristics of variously modified LEP wind turbine blades under various turbulence intensities. J Wind Eng Indust Aerodyn 202:104–188

Autonomous Tilt Rotor Stabilized Plimp Hybrid Airship Unmanned Aerial Vehicle

N. C. Ajay Vishwath, Saras Takearya, Tanishka Mourya, and Ashish Prajapati

1 Literature Review

Moutinho et al. [1] use Lyapunov's theory to analyze nonlinear system stability. Tests have been done to verify the nonlinear performance of controller and thereby correcting the disturbances and errors.

Andan et al. [2] show that the lift force would be three times increased for the airship with wing structure at a positive angle of attack. 20 to 40% of increment in drag occurs with winged airship. Cook et al. [3] explain the various lateral-directional flight modes of the Plimp including modes like sideslip subsidence, yaw subsidence, and oscillatory roll pendulum, also comparison was made between the estimated models and existing airship for various speeds.

DeLaurier et al. [4] performed analysis to develop stability of airships for the non-neutral net buoyancy conditions and non-coincident mass and volumetric centers conditions. Li et al. [5] proposed a method for simulation of airships in nonlinear dynamics. Both the model of the statics and dynamics of air were framed. Wang et al. [6] used CFD and Fourier analysis to obtain the stability derivatives.

Ceruti et al. [7] describe the optimization of airship that consist of two semi-ellipsoids, and axis ratios were altered for the same. The various parameters to optimize were volume, dimension of the tail, ratio between the vertical and the lateral semi-axis, the percentage coverage of photovoltaic films on surface of the top, and the ratio between the longitudinal and the lateral semi-axis. Andan et al. [8] presented the results of a numerical study of aerodynamic parameters for a wingless as well as a winged airship. For various angles, the net force coefficients and moment coefficients have been calculated.

N. C. Ajay Vishwath (✉) · S. Takearya · T. Mourya · A. Prajapati
Parul University, Vadodara 391760, India
e-mail: ajay.nc2934@paruluniversity.ac.in

D. Shukla (ed.), *Lighter Than Air Systems*, Lecture Notes in Mechanical Engineering,
https://doi.org/10.1007/978-981-19-6049-9_15

2 Introduction

The Plimps are the flying devices that can be described as an aircraft with plummet-proof which has lifting capability of an airplane, the thrust control mechanism of a helicopter, and the lift due to buoyancy forces of a Plimp [9]. Egan Airships has designed their eight-passenger Model J. Plimp which are new type of airship, with a combination of helium envelope and dynamic rotors enabled wings, allowing it to perform operations like hovering, dipping, ascending, banking, or spinning [10]. Even at times of engines off mode, it would simply float and glide smoothly to the land. Having VTOL capability, it could lift off from anywhere without runway and land anywhere without runway [11]. The Plimp was initially designed by Daniel P. Raymer, who is a famous expert in the discipline of aircraft conceptual design and aircraft design engineering [12]. Advantage of Plimp being that the aircrafts are noisy, Plimps are not.

3 Design

The pimp would be having a hull, structural frame, thrusters with servos to pivot, the H-tail, and the flight control with an embedded system. The pressurized inflatable envelope would be filled with the suitable gas that is lighter than air that should provide maximum static lift in air for the unmanned aerial airship. Though hull portion of the airship could be shaped with various geometries, the ellipsoidal shaped hull would be preferred for this tilt rotor stabilized hybrid Plimp unmanned aerial vehicle due to high efficiency with less surface area to volume ratio and thereby power consumption could be minimized. This would improve the endurance of the UAV. The tilt rotors would vector the thrust force in necessary directions. Vectoring the thrust components would ease the Plimp to correct its flight path and provide necessary stability in any particular axes. The weight due to gravity, the lift due to aerodynamics, the aerostatic lift, and the thrust vectored lift would be considered as the major force that are acting on this unmanned airship. Among these forces, the aerostatic lift would be given by the differences between the force due to buoyancy and the weight of gas displaced (Fig. 1). Consider volume (V) of the envelope and the density (ρ), the aerostatic lift would be given by

$$\text{Lifting force}\left(L_f\right) = V\left(\rho_{\text{air}} - \rho_{\text{gas}}\right) \tag{1}$$

$$F = m\left(\frac{\mathrm{d}v}{\mathrm{d}t}\right) = \text{Thrust}(F_{\text{t}}) + \text{Buoyancy}(F_{\text{b}}) - \text{Weight}(W)\text{Drag}(F_{\text{d}}) \tag{2}$$

Fig. 1 Free body diagram of takeoff [13]

where

$$\mathrm{Drag}(F_d) = 0.5\, C_D \rho_{air} A v^2 \tag{3}$$

C_D: Drag coefficient and depends on shape,
ρ_{air}: Air density,
A: Area (cross-sectional),
v: Velocity.

The design of envelope has the major impact on the stability, so the necessary design considerations were adopted while designing them. GNVR shape has been chosen to construct the airship. The geometric profile of the airship is given in Fig. 2.

Choice of gas being used I the envelope is also most essential part of the design. So for the comparison, let's tabulate the various gases commonly used along with their densities and molar mass (Table 1).

3.1 Material Selection

Factors affecting the material selection for the envelope; would be the price, the sturdiness, the stress sustaining capability during various flight conditions, including the infiltration of the lifting fluid (gas) [15]. High strength to weight ratio, high tear resistance, resistance to the environmental degradation, and low permeability to LTA gases are the basic material property for choosing such inflatable structure [16, 17]. The biaxial-oriented polyethylene terephthalate in short known as BOPET is selected for inflatable structure material. The metalized BOPET, also known as Mylar, is cheaper than the normal polyurethane. But the studies suggested that it is susceptible to gas (helium) leakage [18]. If the Mylar gets punctured, it would wear out rapidly than the polyurethane material [19].

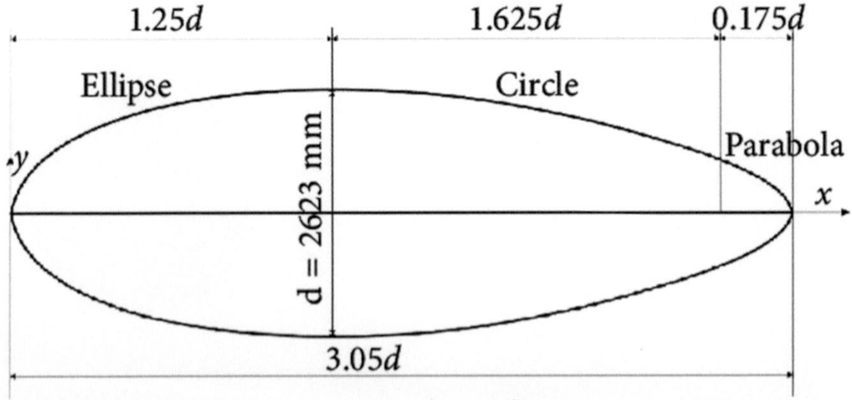

Nose Ellipse - origin to 1.25*d*: $\frac{x^2}{1.25d^2}+\frac{y^2}{0.5d^2}=1$

Middle Circle - 1.25d to 1.62*d*: $x^2+(y-3.5d)^2=16d^2$

Tail Parabola - 1.62d to 1.8*d*: $y^2=1.373d(1.8d-x)$

Fig. 2 GNVR profile for the airship [14]

Table 1 Density and molar mass of gases at standard conditions

Gas used in envelope	Density (kg/m^3)	Molar mass (g/mol)
Helium	0.169	4.0026
Methane	0.73	6.04
Ammonia	0.756	17.031
Hydrogen	0.085	2.02
Air	1.225	28.9647

3.2 *Stability of Airship*

Considering the stability of airships, it could be defined through the classification by static and dynamic. The classification of stability considered during the no powered flight condition would be termed as static stability. The phenomenon of return back to its original position despite of disturbances defines this condition [20] (Table 2).

In general, airships are statically unstable in yaw. Effect of dynamic stability comes in the picture when the airstream flow passes through the control surfaces. Though the stability of airplane and airship seems to have similar classification of stabilities, one of the major differences being that the stabilities in the case of airplanes is associated with one another, but in the case of airships, they being independent of each other. In steady flight, pitch stability, yaw stability, and roll stability are the various stabilities involved.

Table 2 Axes of airship and conventional symbols related to them

Axis		Longitudinal axis	Lateral axis	Normal axis
	Symbol	X	Y	Z
Moment about the axis	Force acting parallel to the axis	X	Y	Z
Angles	Designation	Rolling	Pitching	Yawing
	Symbol	L	M	N
	Positive direction	$Y \rightarrow Z$	$Z \rightarrow X$	$X \rightarrow Y$
Angles	Designation	Roll	Pitch	Yaw
	Symbol	φ	θ	ψ
Velocity components	Linear component about axis	u	V	w
	Angular	p	Q	r

Assumptions that need to be made while performing the derivation of the stability parameters:

(1) The net weight of the body remains constant.
(2) Considering the accessional force to remain constant.
(3) Fixed center of gravity as well as center of buoyancy.
(4) The controls remain in neutral.
(5) Constant velocity.
(6) No changes in the form of airship [20].

3.3 The Various Forces and the Various Moments Acting on the Plimp

Consider Plimp that flying along the horizontal path, such that the flight path makes an angle of 0° with the longitudinal axis, then the various forces and the moments acting on the Plimp would be (see Fig. 3) [20].

(1) Forces

$L_0 =$ Lift of inflating gas acting through center of buoyancy, G.
$W =$ Total weight of dead and live loading, acting through center of gravity, M.
$R =$ Resistance of envelope and appendages, acting through center of pressure, P.
$T =$ Propeller thrust, acting parallel to axis of envelope at distance $o$below$M$.

(2) (2) Moments about M

$$Moment L_0 = L_0 \times 0 = 0.$$
$$Moment W = W \times 0 = 0.$$
$$Moment thrust - resistance couple = T(c + d).$$

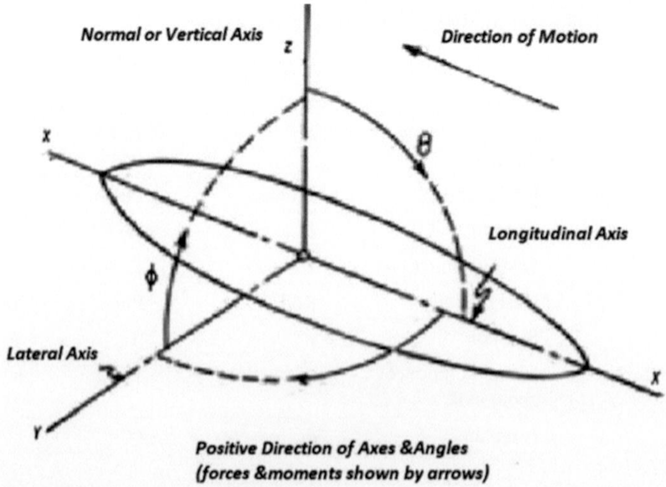

Fig. 3 Axes and angles in positive direction [20]

Condition for static equilibrium and keeping constant velocity

$$L_0 = W \tag{4}$$

$$R = T \tag{5}$$

During the flying condition of Plimp on an even keel, the moments due to thrust force and resistance force would be unbalanced; this would nose up the Plimp. To handle such phenomenon when Plimps are full of gas, are regularly trimmed a few degrees nose heavy. In case of gust, disturbances in the longitudinal axis would give rise to a slight tilt from the horizontal plane, few cases could be observed and described as Table 3.

From Fig. 4, the forces, the lever arms, and the moments, for the cases one to six are noted:

(1) Forces

- L_g = Lifting force of fluid.
- W = The Net Weight acting due to gravity.
- F_e = Resultant air force acting on the envelope.
- L_s = Lift of tail surface.
- F_s = Resultant force acting on the tail surfaces
- T = The thrust of Propeller.
- L_e = Vertical component of the forces due to motion acting on the envelope.
- t = The horizontal component of the thrust produced by propeller .
- R_e = Horizontal component of the forces due to motion acting on the envelope.
- R_s = Drag of the tail surface.

Table 3 Six cases depending on the static state of Plimp with the course of inclination [20]

Case	Static state of airship	Course of inclination
1	Plimp is in the static equilibrium, and nose is tilted up	θ is the angle formed between the direction of motion and longitudinal axis $\theta = 0$ α is the angle between the motion direction and the horizontal, airship climbs at an angle of tilt, α
2	Plimp is in the static equilibrium, and nose is tilted down	$\theta = 0$ and descend angle, α
3	Plimp is statically heavy, and nose is tilted up	Plimp climb at an angle lesser than the angle of tilt, and $(\alpha + \theta)$ will be the angle between longitudinal axis and the horizontal
4	Plimp is statically heavy, and nose is tilted down	$(\alpha - \theta)$ will be the angle between longitudinal axis and horizontal. Plimp descend at an angle greater than the inclination
5	Plimp is statically light, and nose is tilted up	$(\alpha - \theta)$ will be the angle between longitudinal axis and horizontal
6	Plimp is statically light, and nose is tilted down	Plimp descend at an angle lesser than the inclination $(\alpha - \theta)$ will be the angle between longitudinal axis and horizontal

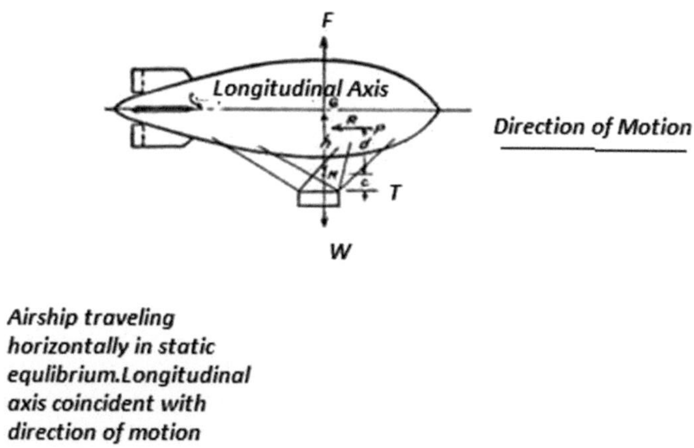

Fig. 4 Forces on airship in horizontal flight [20]

- L_t = The vertical Component of thrust produced by propeller [20].

(2) Leaver Arms

- $W = K \sin(\alpha \pm \theta)$
- $L_g = 0$.

- $T = (c + h)$.
- $F_s = a$(assuming F_s, perpendicular tothe surfaces).
- $L_s = a \times \cos(\alpha \pm \theta)$.
- $R_s = a \times \cos(\alpha \pm \theta)$.
- F_e would vary with the position of P, which in turn would depend on θ.
- $L_e = b \times \cos(\alpha \pm \theta)$.

(3) Moments

- Moment due to weight is $Wh \sin(\alpha \pm \theta)$.
- Moment due to propeller thrust is $T(c + h)$.
- F_e tend to turn the complete Plimp in the positive direction about M due to the increased pressure below the hull. This phenomenon is assisted by reducing the pressure in bottom surface of the tail. The forces acting below the nose of the UAV and the tail of the UAV would be in opposite direction. As the nose force is to some extent is greater than the tail force, there is a difference in force, which will be known as the dynamic lift of hull. Despite of the difference, both the forces have same direction of rotation, and the resultant moment caused is dynamic upsetting moment, denoted by M_e.
- Moment due to the tail surface, denoted by M_s opposes this dynamic upsetting moment. $M_s = L_s\alpha\cos(\alpha \pm \theta) + R_s\alpha \sin(\alpha \pm \theta)$.

3.4 Plimp Stability from Designer Point of View

- Plimps are very stable about their lateral axis.
- Design inputs needed for Plimps to provide longitudinal stability.
- In yaw, Plimps are statically unstable, only pilots can handle this through their rudders.

3.5 Requirements of the Hybrid VTOL Plimp Airship Unmanned Aerial Vehicle

- Payload bay that can carry payload along with the essential electronic components.
- The primary lifting device that would be integrated to the payload bay and installed to provide hydrostatic buoyancy.
- The secondary lifting device integrated to the fuselage and installed to provide dynamic lift through movement of the secondary lifting device through the air.
- The thrust system equipped to generate thrust, the thrust system integrated to the secondary lifting device and it rotates together about an axis that is aligned with the spar of the wing [21].
- The tail system that could be pivoted upon the tail boom to counteract the unwanted forces and moments produced by the tilt rotors [22].

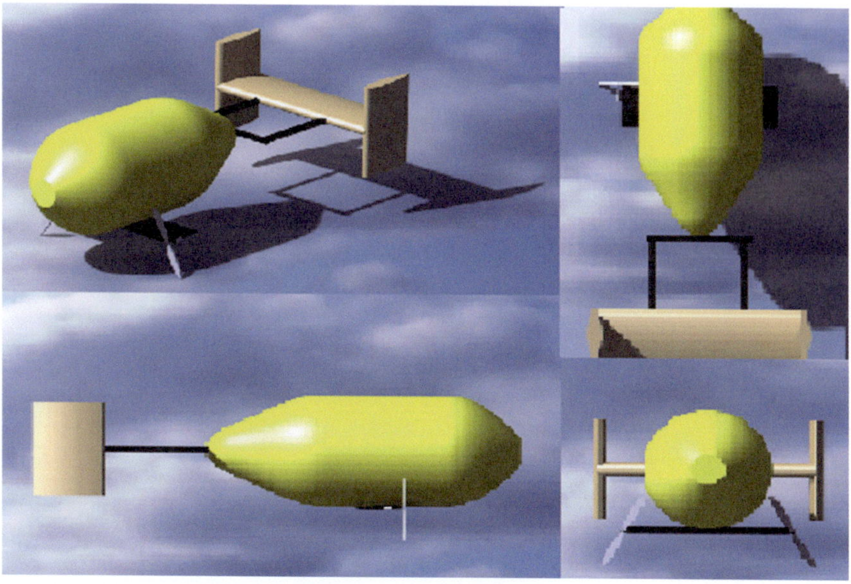

Fig. 5 Isometric, top, side, and front views of Plimp during forward tilt of thrusters

Through the requirements defined above, the hybrid VTOL airship unmanned aerial vehicle with H-tail has been designed using the modeling software CATIA. The various three-dimensional geometric views of the unmanned aerial vehicle are shown in Figs. 5 and 6.

4 Calculations

4.1 Weight Estimation

See Table 4.

4.2 Airfoil Data for the Wing

The airfoil used in this UAV is Bell A821201 (23%) FX-66-H-60, because most of the thrust vector is going to be away from the chord line. This airfoil has flat bottom surface and streamlined upper surface which helps UAV to float stably irrespective of thrust direction.

Fig. 6 Isometric, top, side, and front views of Plimp during upward tilt of thrusters

Table 4 Weight estimation of the UAV

Sr no.	Parts	Weight (gm)
1	Envelope with gas	1800
2	BLDC motors	320
3	ESC (2)	80
4	Propellers (2)	40
5	Battery	250
6	Carbon fibertail boom	300
7	Elevator	100
8	Fuselage	200
9	Servo motor (3)	60
10	Receiver	20
11	Extras	130
	Total weight	3300

4.3 Propeller Data

Propeller: Radius 12.7 cm.

$$\text{Area} = \pi r^2 = 0.507\,\text{m}^2 \tag{6}$$

4.4 *Different Phases of Flight*

Hovering

Calculations of various performance parameters during hovering are given by [23].

Generally, the thrust produced by the motors and the propellers combination should be sufficient to lift the total weight including payload of the UAV

$$T = 2\rho A(V + V_i)V_i = 3.3\,\text{kg} = 32.34\,\text{N} \tag{7}$$

$$V_\text{h} = \sqrt{\frac{W}{2\rho A}} = 16.14\,\text{m/s} \tag{8}$$

$$\Omega = \frac{2\pi N}{60} = 1278.62\,\text{rad/s} \tag{9}$$

$$V_\text{h} = \Omega R\sqrt{\frac{C_T}{2}} \tag{10}$$

Thrust of a single BLDC motor with the propeller attached is given by

$$T = C_\text{T}\rho(\Omega R)^2\ A = 32.4265\,\text{N} \tag{11}$$

Torque of a single BLDC motor with the propeller attached is given by

$$Q = C_Q\rho(\Omega R)^2\ AR = 0.5122\,\text{Nm} \tag{12}$$

$$\text{F.O.M} = \frac{C_\text{T}^{3/2}}{\sqrt{2}C_\text{Q}} \tag{13}$$

$$P = Q\ \Omega = 654.89\,\text{Nm/s} \tag{14}$$

Climbing

Calculations of various performance parameters during climbing are given by [23]

$$D = 0.5\,\rho(V + V_i)^2 A_\text{B} C_\text{DB} = 8.9484\,\text{N} \tag{15}$$

$$T = D + W = 41.2936\,\text{N} \tag{16}$$

Therefore,

$$\rho A(V + V_i)2V_i = 0.5\rho(V + V_i)^2 A_{\mathrm{B}} C_{\mathrm{DB}} + W \tag{17}$$

$$Q = 4A(V + V_i)\ \omega R^2 = 1.02097\,\mathrm{Nm} \tag{18}$$

$$Q(\Omega - \omega) = T(V + V_i) \tag{19}$$

$$P = Q\ \Omega = 1305.4327\,\mathrm{Nm} \tag{20}$$

where

V_c is climbing velocity,
D: Drag generated,
AB is the area of the propeller,
C_{D_B} is the drag coefficient due to the propeller.

Forward

Calculations of various performance parameters during forward are given by [23]

$$D = 0.5\ \rho V_{\mathrm{R}}^2 A_{\mathrm{B}} C_{\mathrm{DB}} = 34.7802\,\mathrm{N} \tag{21}$$

$$\tan\delta = \frac{D\cos\varepsilon}{D\sin\varepsilon + W},$$
$$\delta = 46.5301 \tag{22}$$

$$T^2 = D^2 + W^2 + DW\ \sin\varepsilon = 47.822 \tag{23}$$

$$T = 2A\rho V_i\ V_{\mathrm{R}} = 19.2497 \tag{24}$$

$$\mathrm{Q}\ =\ \mathrm{A}\rho V_{\mathrm{R}} R^2\ \omega = 5.1341\,\mathrm{Nm} \tag{25}$$

$$P = Q\,\Omega = 6564.5629\,\mathrm{Nm} \tag{26}$$

5 Mathematical Modeling and Autopilot Control System

The autopilot control system could be developed through modeling the necessary equations that need to be damped from the kinematics of flight [24].

The equations of motion for damping the pitching moments

$$\sum \text{Pitching moments} = \sum M_{\text{cg}} = I_{y\ddot{\theta}} \tag{27}$$

The pitching moment denoted by M and the pitching angle denoted by θ.

M and θ in terms of the initial reference value are mentioned with subscript 0, and the corresponding perturbation is mentioned by Δ

$$M = M_0 + \Delta M \tag{28}$$

$$\theta = \theta_0 + \Delta\theta \tag{29}$$

If case that the reference moment which is denoted by M_0 becomes 0, then the Eq. (27) reduces to

$$\Delta M = I_y \Delta\ddot{\theta} \tag{30}$$

where

$$\Delta M = \frac{\partial M}{\partial \alpha}\Delta\alpha + \frac{\delta M}{\delta \dot{\alpha}}\Delta\dot{\alpha} + \frac{\partial M}{\partial q}\Delta q + \frac{\partial M}{\partial \delta_e}\Delta\delta_e \tag{31}$$

As there is a constrain applied to the C.G, the angle of attack will be identical to the pitch angle

$$\Delta\alpha = \Delta\theta \tag{32}$$

$$\Delta\dot{\theta} = \Delta\dot{\alpha} \tag{33}$$

$$\dot{\theta} = \Delta q \tag{34}$$

After substitution of the expression into Eq. (30), thereby rearranging would yield.

$$\Delta\ddot{\alpha} - \left(M_q - M_{\dot{\alpha}}\right)\Delta\dot{\alpha} + M_\alpha \Delta\alpha = M_{\delta_e}\Delta\delta_e \tag{35}$$

$$M_q = \frac{\partial M/\partial q}{I_y} \tag{36}$$

For the Plimp, the term M_α is negligible and could be eliminated in calculations.

Characteristics equation for Eq. (34) is

$$\lambda^2 - \left(M_q + M_{\dot{\alpha}}\right)\lambda - M_\alpha = 0. \tag{37}$$

The undamped natural frequency ω_n of the system and damping ratio ζ can be determined by

Table 5 Result table of various parameters

Hovering conditions	Climbing conditions	Forward conditions
$V_h = 11.42$ m/s	$V_i = 10.795$ m/s	$V_i = 19.2497$ m/s
$T = 32.34$ N	$T = 41.2936$ N	$T = 47.822$ N
$D =$ Neglected	$D = 8.9484$ N	$D = 34.7802$ N
$Q = 0.5122$ Nm	$Q = 1.02097$ Nm	$Q = 5.1341$ Nm
$P = 654.89$ Nm/s	$P = 1305.4327$ Nm/s	$P = 6564.5629$ Nm/s

$$\omega_n = \sqrt{-M_\alpha} \tag{38}$$

$$\zeta = -\frac{\left(M_q + M_{\dot{\alpha}}\right)}{2\sqrt{-M_\alpha}} \tag{39}$$

For a step change in rudder control, the solution to Eq. (35) would yield a damped sinusoidal motion, considering that the Plimp UAV has enough aerodynamic damping.

6 Results

We have obtained necessary parameters in hovering conditions, climbing conditions, and forward conditions are found out to be Table 5.

When we feed the rudder transfer function in the aircraft transfer function block in the damper block diagram and giving rudder servo equation as $\frac{10}{(S+10)}$.

Washout circuit equation is given by $\frac{s}{s+\frac{1}{\tau}}$ and S (yrg) is given as $1.04\frac{v}{deg/_{sec}}$.

After obtaining the final equation through MATLAB, we use SISO tool toolbar to find the individual root locus and the final stability could be checked through the graph obtained and by varying the gain (Figs. 7, 8, 9, 10, 11, 12 and 13).

7 Summary

A detailed explanation of various design criteria of Plimp hybrid airship unmanned aerial vehicle has discussed along with their definitions and design constraints. Through the inputs of various stability parameters, necessary requirements of the hybrid VTOL Plimp airship have been defined in Sect. 3. The design of the Plimp unmanned aerial vehicle was modeled in CATIA software to get the exact geometric parameters. The calculations for performance parameters after weight estimation were performed in Sect. 4. The detailed calculations for different phases of flight were demonstrated. Further in Sect. 4, a detailed methodology of mathematical

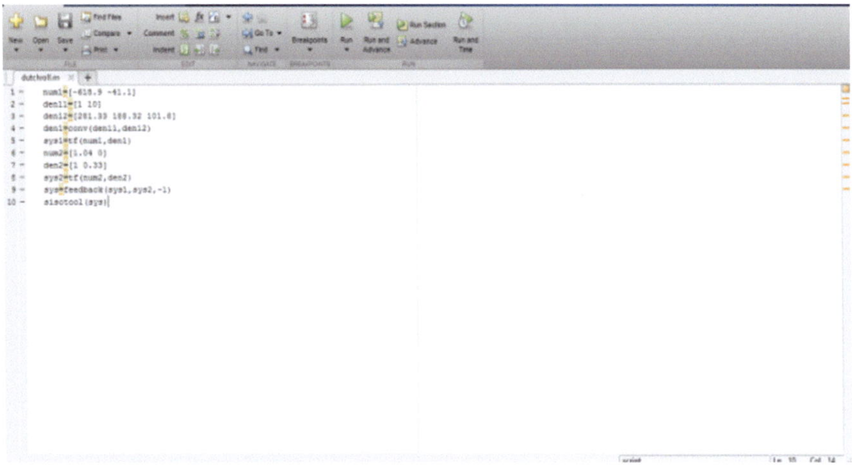

Fig. 7 Program for solving the transfer function MATLAB

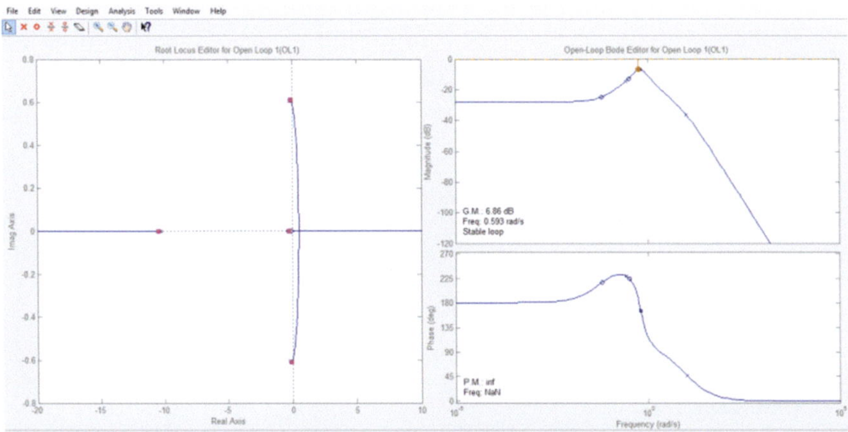

Fig. 8 Graph of the solution for the transfer function in MATLAB

modeling of the stability parameters was derived and necessary equations that would be required for MATLAB code was obtained. In Sect. 5, the coding was performed in MATLAB SISO toolbox and through tuning various gain values, the designed system becomes stable as shown in the graphs of Sect. 6.

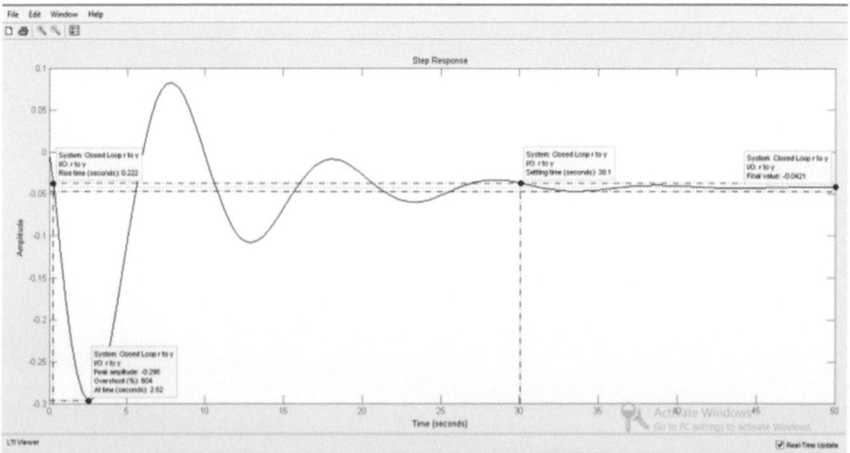

Fig. 9 Step response when gain $k = 1$, damping ratio = 0.203, natural frequency = 0.923, stable loop [8]

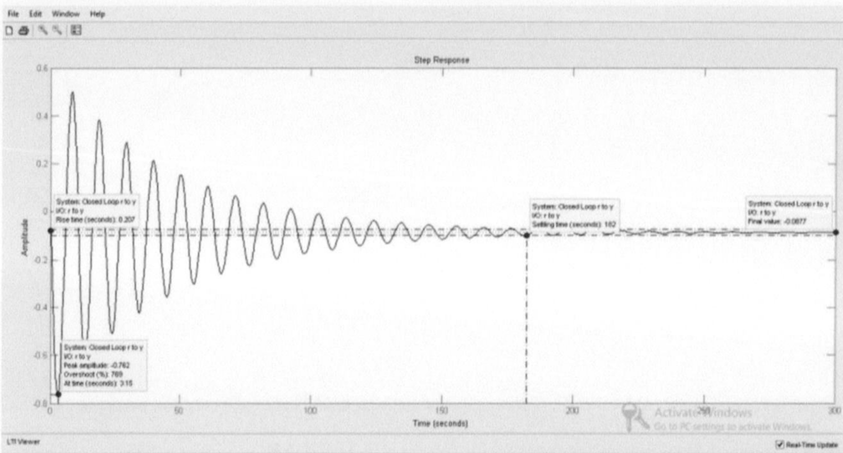

Fig. 10 Step response when gain $k = 3$, damping ratio = 0.036, natural frequency = 0.868, stable loop

8 Conclusion

The designed tilt rotor stabilized Plimp unmanned aerial vehicle produce enough thrust, torque, and power with least possible drag. Also the MATLAB results show the stability for various gain values with different natural frequencies at various damping ratio. From the graph obtained in MATLAB SISO toolbox, we can see that the disturbances get damped and the system become stable. So this manuscript concludes

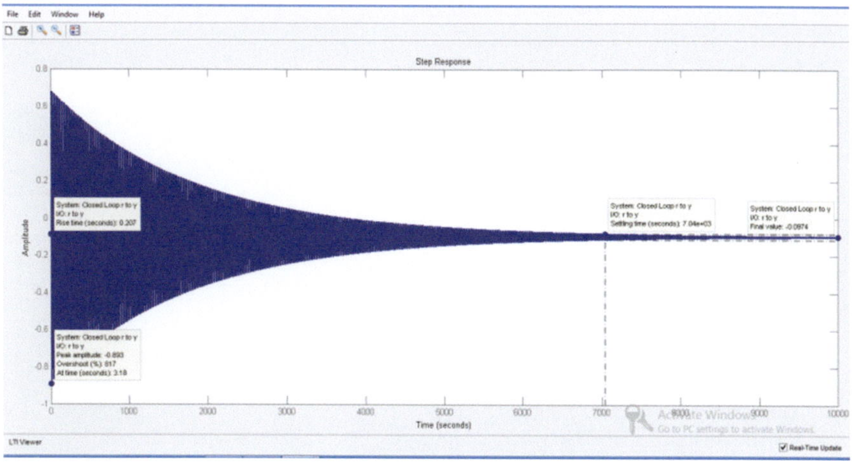

Fig. 11 Step response when gain $k = 3.3$, damping ratio = 0, natural frequency = 0.865, stable loop

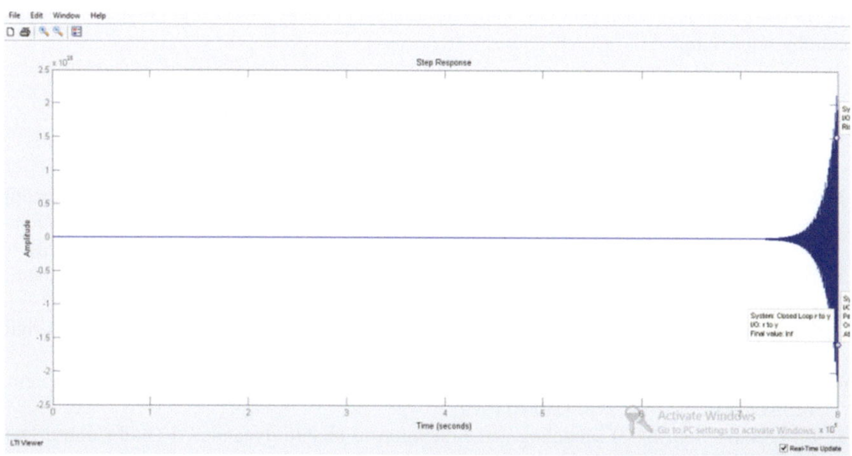

Fig. 12 Step response when gain $k = 3.3$, damping ratio = −0, natural frequency = 0.864, unstable loop

that the modeled autopilot control system stabilizes the disturbances produced in Plimp hybrid airship UAV.

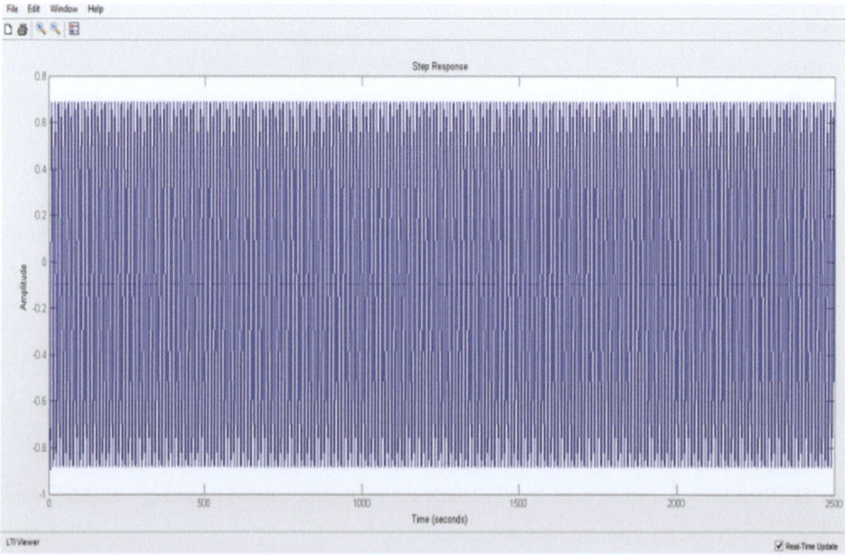

Fig. 13 Step response when gain $k = 3.302651$, damping ratio $= -0$, natural frequency $= 0.864$, neutrally stable loop

References

1. Moutinho Ass, & Azinheira, J. R. Stability and robustness analysis of the AURORA airship control system using dynamic inversion. In Proceedings of the 2005 IEEE International Conference on Robotics and Automation (pp. 2265–2270). Barcelona, Spain(2005, April)
2. Andan AD, Asrar W, Omar AA (2012) Investigation of aerodynamic parameters of a hybrid airship. J Aircr 49(2):658–662
3. Cook MV, Lipscombe JM, Goineau F (2000) Analysis of the stability modes of the non-rigid airship. The aeronautical journal 104(1036):279–290
4. DeLaurier, J., & Schenck, D.). Airship dynamic stability. In 3rd Lighter-Than-Air Systems Technology Conference (p. 1591). (1979)
5. Li Y, Nahon M (2007) Modeling and simulation of airship dynamics. J Guid Control Dyn 30(6):1691–1700
6. Wang XL (2012) Computational fluid dynamics predictions of stability derivatives for airship. J Aircr 49(3):933–940
7. Ceruti A, Voloshin V, Marzocca P (2014) Heusristic algorithms applied to multidisciplinary design optimization of unconventional airship configuration. J Aircr 51(6):1758–1772
8. Andan AD, Asrar W, Omar AA (2012) Aerodynamics of a hybrid airship. In: AIP conference proceedings. Am Inst Phys 1440(1):154–161
9. Hybrid vtol vehicle Homepage. https://patents.google.com/patent/US20160137281/. Last accessed 30 Mar 2022
10. The Plimp airship Homepage. https://www.wearefinn.com/topics/posts/the-plimp-airship-aeroplane-hybrid-is-now-available-to-pre-order/. Last accessed 30 Mar 2022
11. Egan Airships Homepage. https://lynceans.org/wp-content/uploads/2021/04/Egan-Airships_PLIMP_R1-converted.pdf/. Last accessed 30 Mar 2022
12. Dronerush Homepage. https://dronerush.com/blimps-flying-advertisement-drones-10660/. Last accessed 30 Mar 2022

13. Calculating Lifting Capacity of Airships Homepage. https://erik-engheim.medium.com/calculating-lifting-capacity-of-airships-48df5cd7d147/. Last accessed 30 Mar 2022
14. Pant RS (2014) Design, fabrication and flight demonstration of a remotely controlled airship for snow scientists. J Aeros Technol Manag 6:19–27
15. Sadasivan N (2019) Design and realization of an unmanned aerial rotorcraft vehicle using pressurized inflatable structure. Int J Aviat Aeronaut Aeros 6(4):3
16. Miller J, Nahon M (2005) The design of robust helium aerostats. In: AIAA 5th ATIO and 16th lighter-than-air sys technology, and balloon systems conferences. Arlington, Virginia, p 7441
17. Khoury GA (ed) (2012) Airship technology. Cambridge University Press, p 10
18. Nordestgaard M, Ravenscroft L, Bartel N (2007) Design and build a small airship
19. Boon NK (2004) Mini airship patrol craft. National University of Singapore
20. Technical Manual of Airship Aerodynamics Homepage, https://www.faa.gov/regulations_policies/handbooks_manuals/aviation/media/airship_aerodynamics.pdf/. Last accessed 30 Mar 2022
21. Wang X, Cai L (2015) Mathematical modeling and control of a tilt-rotor aircraft
22. Johnson W (1984) Ames research center moffett field, an assessment of the capability to calculate tilting prop-rotor aircraft performance, load and stability, NASA Technical Paper 2291, California
23. Seddon J (1990) PhD, DSc, CEng, CFF, FRA. Blackwell Publications Ltd., Basic Helicopter Aerodynamics
24. Nelson RC (1998) Flight stability and automatic control, 2nd ed. WCB/McGraw Hill, New York

Numerical Approach to Maneuver Design and Feasibility Evaluation for the Autonomy of Airship

Duraisamy Gobiha and Nandan K. Sinha

1 Introduction

Feasible or implementable maneuver design and evaluation is an indispensable area of research in the aerospace control fraternity. Literature rarely emphasizes the feasibility of the chosen desired trajectory or path. Among the ones that contemplate feasibility, optimization is the most prevalent approach, with one being an opportune byproduct of the other [1]. Optimization-based control design like model predictive control (MPC) when implemented on real-time systems requires strictly guaranteed convergence and heavy computational resources mainly for high-performance aerospace applications. Besides, such integration of optimization and control questions the reliability of controls in real-time implementation with modeled and unmodeled uncertainties, process, measurement noise, etc. [1].

Most practical applications merely need a feasible solution rather than an optimal one. The recent introduction of computational guidance and control [1, 2] further provokes the importance of an alternate methodology to obtain feasible solutions. But there are very few non-optimal feasible maneuver design techniques discussed in the literature. Trajectory generation based on analytically solving trim solutions is proposed in [3]. Besides being cumbersome, this method does not give explicit solutions and does not contemplate on stability. In [4], appropriate control inputs pertaining to the reference trajectory are obtained by solving the system's dynamical equations. This is just a validation procedure and relies on trial-and-error for obtaining a feasible reference trajectory.

Maneuver design based on bifurcation analysis has been attempted in [5] using continuation algorithm, AUTO2000. This work generates feasible solutions and uses two trim points between which state variables are switched, to achieve the maneuver.

D. Gobiha (✉) · N. K. Sinha
Department of Aerospace Engineering, Indian Institute of Technology Madras, Chennai 600036, India
e-mail: gobiha@yahoo.com

D. Shukla (ed.), *Lighter Than Air Systems*, Lecture Notes in Mechanical Engineering,
https://doi.org/10.1007/978-981-19-6049-9_16

This method does not perceive trajectory design as a whole and continuation algorithms using AUTO2000 are highly dependent on the initial conditions and are computationally demanding. This significantly limits the applicability of this methodology. But with the recent development of MATLAB® embedded continuation algorithms like MATCONT [6], a direct continuation methodology for constrained system analysis has been proposed which significantly overcomes the complexity of the continuation procedure [7].

This work uses the bifurcation-based continuation technique to compute feasible solutions for different maneuvers regarded for autonomous operations of stratospheric airship. The computed solutions take into account the complete dynamics of the considered airship model with its state and control constraints. Some of the challenging aspects of stratospheric airship like ascending, descending, and hovering with a minimal lateral excursion are interpreted and validated.

The paper is structured as follows. Section 2 presents the mathematical modeling of the considered six degrees-of-freedom stratospheric airship model. Section 3 elaborates the numerical description of bifurcation analysis for multi-input multi-output dynamical systems. Maneuver design for various flight conditions like a level straight, level turn, hover, ascend, and descend is discussed in Sect. 4. Simulation for the validation of formulated maneuvers is carried out in Sect. 5. Section 6 concludes the paper.

2 6-DoF Aircraft Equations of Motion

This work considers an airship at an operational altitude of 21 km. Such airships are called high altitude or stratospheric airships as they operate in the stratospheric regime of the atmosphere. Airships in this regime face minimal intrusions from other aerial vehicles and provide better remote sensing due to their proximity to the ground. Reusability along with its greener, cheaper and quieter operations, makes stratospheric airship an effective platform for surveillance, atmospheric measurements, disaster management, and space tourism. Though real-time implementation of low altitude airships is uncomplicated, mainly in ascend and descend phases, stratospheric airships have strategic advantages like low magnitude wind speed as inferred from the global atmospheric wind profile [8] and extensive geographic area coverage, which serves enormous applications.

The considered airship is Gertler shaped with a hull filled with helium gas and ballonets filled with air to regulate the internal pressure of airship. A cruciform tail configuration at the rear serves as the control actuator for the airship. The fin deflections of four arms are characterized as elevator (δ_e), aileron (δ_a), and rudder (δ_r) angles as illustrated in Fig. 1 for the ease of incorporation in aerodynamic model of the airship. Resultant of fin deflections that generates pitch, yaw, and roll forces and moments are defined as elevator (δ_e), aileron (δ_a), and rudder (δ_r) angles as tabulated in Eqs. (1)–(3),

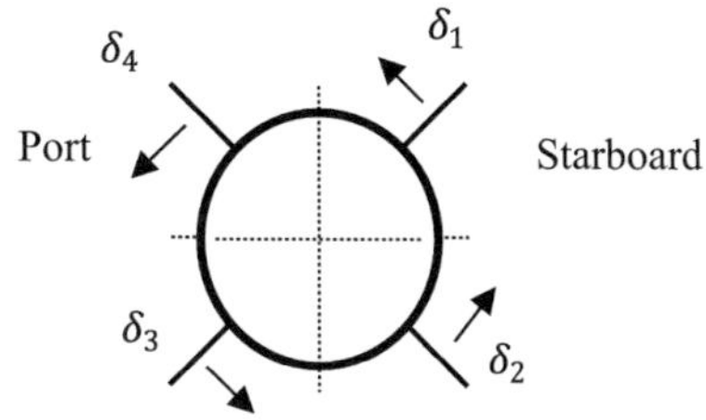

Fig. 1 Rear view of tail configuration

$$\delta_e = \frac{\delta_1 + \delta_2 - \delta_3 - \delta_4}{4} \tag{1}$$

$$\delta_r = \frac{-\delta_1 + \delta_2 + \delta_3 - \delta_4}{4} \tag{2}$$

$$\delta_a = \frac{\delta_1 + \delta_2 + \delta_3 + \delta_4}{4} \tag{3}$$

Structural analysis, stability, and control derivatives and inertial values of the airship were extensively evaluated in [9]. Six degrees-of-freedom modeling of airship should consider inertial, aerodynamic, gravitational, buoyant, and propulsive forces. In this work, equations of motion of airship are derived with components along the wind axes frame with origin, o at the center of volume (CV). Airship equations of motions are similar to that of aircraft except for the influence of buoyant force on translational and rotational dynamic equations. Buoyant force along with virtual mass and inertia terms has a significant influence on the dynamic equations, whereas translational and rotational kinematic equations are the same as that of aircraft, as tabulated in Eqs. (4)–(9) [10]

$$\begin{aligned}\dot{x}_E = {} & V\cos\alpha\cos\beta(\cos\psi\cos\theta) \\ & + V\sin\beta(\cos\psi\sin\theta\sin\phi - \sin\psi\cos\phi) \\ & + V\sin\alpha\cos\beta(\cos\psi\sin\theta\cos\phi + \sin\psi\sin\phi)\end{aligned} \tag{4}$$

$$\begin{aligned}\dot{y}_E = {} & V\cos\alpha\cos\beta(\sin\psi\cos\theta) \\ & + V\sin\beta(\sin\psi\sin\theta\sin\phi + \cos\psi\cos\phi) \\ & + V\sin\alpha\cos\beta(\sin\psi\sin\theta\cos\phi - \cos\psi\sin\phi)\end{aligned} \tag{5}$$

$$\begin{aligned}\dot{z}_E = {} & V\cos\alpha\cos\beta(-\sin\theta) + V\sin\beta(\cos\theta\sin\phi) \\ & + V\sin\alpha\cos\beta(\cos\theta\cos\phi)\end{aligned} \tag{6}$$

$$\dot{\phi} = p + q\sin\phi\tan\theta + r\cos\phi\tan\theta \tag{7}$$

$$\dot{\theta} = q\cos\phi - r\sin\phi \tag{8}$$

$$\dot{\psi} = \sec\theta(q\sin\phi + r\cos\phi) \tag{9}$$

2.1 Buoyant and Gravitational Terms

The buoyant force acts along $-Z_E$ axis, whereas gravitational force acts along $+Z_E$ axis. On using the transformation matrix as contemplated in [10], buoyant and gravitational forces $(\boldsymbol{F}_{bg})$ are transformed from inertial axes system to wind axes system.

$$\boldsymbol{F}_{bg} = \begin{bmatrix} 0 \\ 0 \\ mg - B \end{bmatrix}_E = \begin{bmatrix} -(mg - B)\sin\gamma \\ (mg - B)\cos\gamma\sin\mu \\ (mg - B)\cos\gamma\cos\mu \end{bmatrix}_W \tag{10}$$

Moment generated on the airship due to buoyant and gravitational terms $(\boldsymbol{M}_{bg})$ is illustrated in [8].

$$\boldsymbol{M}_{bg} = \begin{bmatrix} -(mga_z + Bb_z)\sin\phi\cos\theta \\ -(mga_z + Bb_z)\sin\theta - (mga_x + Bb_x)\cos\phi\cos\theta \\ (mga_x + Bb_x)\sin\phi\cos\theta \end{bmatrix} \tag{11}$$

2.2 Aerodynamic and Control Terms

Aerodynamic and control force and moment act along the stability axes of the system. Thus on transforming it in wind axis frame, aerodynamic force $(\boldsymbol{F}_{\mathrm{a}})$ is given by

$$\boldsymbol{F}_{\mathrm{a}} = \begin{bmatrix} -D \\ Y \\ -L \end{bmatrix}_S = \frac{1}{2}\rho V^2 S \begin{bmatrix} -C_D\cos\beta + C_Y\sin\beta \\ C_D\sin\beta + C_Y\cos\beta \\ -C_L \end{bmatrix}_W \tag{12}$$

Similarly, aerodynamic moment vector $(\boldsymbol{M}_{\mathrm{a}})$ is given by

$$\boldsymbol{M}_{\mathrm{a}} = \begin{bmatrix} \mathcal{L} \\ M \\ N \end{bmatrix} = \frac{1}{2}\rho V^2 S \begin{bmatrix} bC_l \\ cC_m \\ bC_n \end{bmatrix} \tag{13}$$

2.3 Propulsive Terms

The propulsive system consists of a symmetric pair of thrusters on the aft of CV. The thruster on the port and starboard sides of airship is represented by T_{p} and T_{s}, and they are inclined at an angle ζ_{p} and ζ_{s}, respectively. In this work, it is assumed that the direction of tilt angles and magnitude of thrust along both port and starboard sides are identical. The coordinates of both thrusters are (d_x, d_y, d_z) as shown in Fig. 2.

The propulsive force along body axes frame is

$$\boldsymbol{F}_{\mathrm{p}} = \begin{bmatrix} T_{\mathrm{s}} \cos \zeta_{\mathrm{s}} + T_{\mathrm{p}} \cos \zeta_{\mathrm{p}} \\ 0 \\ -(T_{\mathrm{s}} \sin \zeta_{\mathrm{s}} + T_{\mathrm{p}} \sin \zeta_{\mathrm{p}}) \end{bmatrix}_B = \begin{bmatrix} (T_{\mathrm{s}} + T_{\mathrm{p}}) \cos \zeta \\ 0 \\ -(T_{\mathrm{s}} + T_{\mathrm{p}}) \sin \zeta \end{bmatrix}_B \tag{14}$$

On transforming the force from body to wind axes system, Eq. (14) becomes

$$\boldsymbol{F}_{\mathrm{p}} = \begin{bmatrix} (T_{\mathrm{s}} + T_{\mathrm{p}}) \cos \zeta \cos \alpha \cos \beta - (T_{\mathrm{s}} + T_{\mathrm{p}}) \sin \zeta \sin \alpha \cos \beta \\ -(T_{\mathrm{s}} + T_{\mathrm{p}}) \cos \zeta \cos \alpha \sin \beta + (T_{\mathrm{s}} + T_{\mathrm{p}}) \sin \zeta \sin \alpha \sin \beta \\ -(T_{\mathrm{s}} + T_{\mathrm{p}}) \cos \zeta \sin \alpha - (T_{\mathrm{s}} + T_{\mathrm{p}}) \sin \zeta \cos \alpha \end{bmatrix}_W \tag{15}$$

Similarly, moment due to propulsive system $(\boldsymbol{M}_{\mathrm{p}})$ is given by

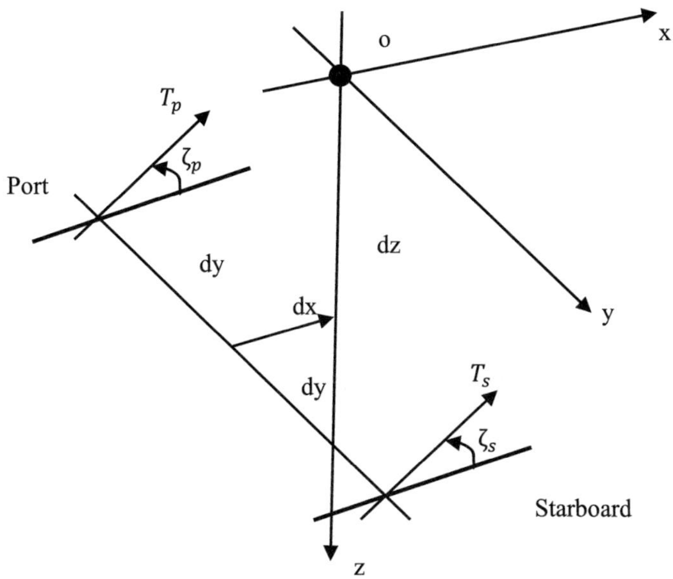

Fig. 2 Geometry of propulsive system

$$\boldsymbol{M}_{\mathrm{p}} = \begin{bmatrix} (T_{\mathrm{p}} \sin \zeta_{\mathrm{p}} - T_{\mathrm{s}} \sin \zeta_{\mathrm{s}}) d_y \\ T_{\mathrm{p}}(d_z \cos \zeta_{\mathrm{p}} - d_x \sin \zeta_{\mathrm{p}}) + T_{\mathrm{s}}(d_z \cos \zeta_{\mathrm{s}} - d_x \sin \zeta_{\mathrm{s}}) \\ (T_{\mathrm{p}} \cos \zeta_{\mathrm{p}} - T_{\mathrm{s}} \cos \zeta_{\mathrm{s}}) d_y \end{bmatrix}$$

$$= \begin{bmatrix} (T_{\mathrm{p}} - T_{\mathrm{s}}) \sin \zeta d_y \\ (T_{\mathrm{p}} + T_{\mathrm{s}})(d_z \cos \zeta - d_x \sin \zeta) \\ (T_{\mathrm{p}} - T_{\mathrm{s}}) \cos \zeta d_y \end{bmatrix} \tag{16}$$

Assumptions that are incorporated based on design aspects of the airship are

(1) Symmetric about xz plane with both center of buoyancy (CB) and center of gravity (CG) in that plane, i.e., $a_y = b_y = 0$.
(2) CV coincides with CB, i.e., $a_x = a_z = 0$ and CB is above CV, i.e.,$b_x = 0$.
(3) Thrusters are assumed to be positioned below CV, therefore, $d_x = 0$.
(4) This work also assumes non-differential thrust and angle, i.e., $T_{\mathrm{s}} = T_{\mathrm{p}} = T/2$ and $\zeta_{\mathrm{s}} = \zeta_{\mathrm{p}} = \zeta$.
(5) Neutral buoyancy is considered throughout the entire flight regime.
(6) Airship is considered an open system, with mass remaining constant and volume varying with the inflation and deflation of ballonets.

Based on these assumptions and the derived terms, Newton's second law of motion is applied to each degree-of-freedom [8]. Translational and rotational dynamic equations are subsequently arrived at and are presented in Eqs. (17)–(22).

$$\dot{p} = \left(\frac{J_y - J_z}{J_x}\right) qr + \left(\frac{J_{xz}}{J_x}\right) pq - \left(\frac{Bb_z}{J_x}\right) \sin \phi \cos \theta + \frac{1}{2J_x} \rho V^2 SbC_l \tag{17}$$

$$\dot{q} = \left(\frac{J_z - J_x}{J_y}\right) pr + \left(\frac{J_{xz}}{J_y}\right)(r^2 - p^2) - \left(\frac{Bb_z}{J_y}\right) \sin \theta + \left(\frac{T}{J_y}\right) d_z \cos \zeta + \frac{1}{2J_y} \rho V^2 ScC_m \tag{18}$$

$$\dot{r} = \left(\frac{J_x - J_y}{J_z}\right) pq - \left(\frac{J_{xz}}{J_z}\right) qr + \frac{1}{2J_z} \rho V^2 SbC_n \tag{19}$$

$$\dot{V} = \frac{1}{m_x}\left(T \cos \beta (\cos \zeta \cos \alpha - \sin \zeta \sin \alpha) - \frac{1}{2} \rho V^2 S (C_D \cos \beta - C_Y \sin \beta) - (mg - B) \sin \gamma \right) \tag{20}$$

$$\dot{\alpha} = q - \frac{1}{\cos \beta}\left\{ (p \cos \alpha + r \sin \alpha) \sin \beta + \frac{1}{V m_z}\left(T \cos \zeta \sin \alpha + T \sin \zeta \cos \alpha + \frac{1}{2} \rho V^2 S C_L - (mg - B) \cos \mu \cos \gamma \right)\right\} \tag{21}$$

$$\dot{\beta} = (p \sin\alpha - r \cos\alpha) + \frac{1}{Vm_y}(-T \sin\beta(\cos\zeta \cos\alpha - \sin\zeta \sin\alpha)$$
$$+ \frac{1}{2}\rho V^2 S(C_Y \cos\beta + C_D \sin\beta) + (mg - B) \sin\mu \cos\gamma\Big) \quad (22)$$

3 Direct Continuation Methodology for Constrained Dynamical Systems

The dynamics of most real-world systems could be captured by a set of ordinary nonlinear first order differential equations of form

$$\dot{\boldsymbol{x}} = \boldsymbol{f}(\boldsymbol{x}, \boldsymbol{u}) \quad (23)$$

where $\boldsymbol{x} \in \mathbb{R}^n$ is the state vector of n variables and $\boldsymbol{u} \in \mathbb{R}^m$ is the control vector of m variables. For a six degrees-of-freedom aerial vehicles, state and control vectors are given by $\boldsymbol{x} = [V, \alpha, \beta, p, q, r, \phi, \theta]^T \in \mathbb{R}^8$ and $\boldsymbol{u} = [\eta, \delta_e, \delta_a, \delta_r]^T \in \mathbb{R}^4$, respectively. Dynamical stability evaluation of such a multi-input multi-output system using time simulation is ineffective. It would involve numerous combinations of control parameters with an infinite loop of initial condition dependency. Numerical continuation-based bifurcation methodology is an effective alternative to analyze such systems. This approach computes a series of steady states called equilibrium or trim solutions by simultaneously solving the algebraic equations for each control parameter by setting $\dot{x} = 0$ in Eq. (23). Stability of the system at each trim solution is also interpreted by calculating eigenvalues from its corresponding Jacobian matrix. Thus, characterizing the system locally gives a picture of global dynamics of the system.

The continuation approach solves Eq. (23) as a function of a single control parameter called continuation parameter, while retaining all other control parameters at a fixed value. The remaining control parameters called free variables could be tactically used to satisfy a set of constraints that defines a flight condition or maneuver. This is illustrated in Eq. (24), where $s \in \boldsymbol{u}$ is the continuation parameter and $\boldsymbol{p} \in \boldsymbol{u} \in \mathbb{R}^3$ are free variables.

$$\dot{\boldsymbol{x}} = \boldsymbol{f}(\boldsymbol{x}, s, \boldsymbol{p}) \quad (24)$$

This results in control schedules with respect to continuation parameter to satisfy the imposed constraints. It also helps in gauging the achievability of imposed constraints with the available control effort. It is notable that the number of constraint equations should not exceed $m - 1$ and free variables should have a significant influence on the constraint variables. Such constrained system analysis helps in the performance evaluation of maneuvers and in feasible maneuver design [11, 12]. A recent development of direct continuation methodology for constrained system analysis [7] using MATCONT toolbox in MATLAB® substantially reduces the computational complexity and initial condition dependency. Thus, this methodology has been utilized to extend the capabilities of bifurcation-based continuation technique to trajectory design, feasibility evaluation, and optimization.

4 Maneuver Design with Feasibility Evaluation

This section evaluates different maneuvers for the autonomous operation of airship. At the operational altitude, airships may need to follow a straight level flight or exhibit turn maneuvers. Gauging the performance of turn maneuvers also helps in serving some of the major functionalities of airship like station-keeping. Besides analyzing the performance of airship at its pressure altitude, it is also important to assess its behavior during ascend and descend phases which are the challenging aspects of stratospheric airship platform. This section, therefore, evaluates the performance of airship in level, turn, ascend, and descend maneuvers. This section assumes neutral buoyancy and zero tilt angle for thrusters in level flight conditions.

4.1 Straight and Level Flight Maneuver

This section generates successive trim solutions for straight and level flight condition. These trims correspond to airship flying in straight line at pressure altitude with zero sideslip and wings level conditions. These conditions transcribe as constraint equations specified in Eq. (25).

$$\gamma = 0; \beta = 0; \phi = 0 \tag{25}$$

These three constraints are satisfied by freeing three control variables. Control variables to be freed should be appropriately chosen such that they have substantial effect on constraints. In this case, η, δ_r, and δ_a are freed to satisfy constraints on γ, β, and ϕ, respectively. In the bifurcation plots in Fig. 3, solid lines represent stable trim solutions and dotted lines correspond to unstable trim solutions. Thus, it could be inferred that Hopf bifurcation occurs at the state and control trim vectors specified in Eqs. (26) and (27), respectively, with velocity in m/s and angles in degrees. The instability is caused due to divergence of pendulum mode at higher velocities.

$$\boldsymbol{x} = [15.7686, -0.3912, 0, 0, 0, 0, 0, -0.3912]^{\mathrm{T}} \tag{26}$$

$$\eta = 0.2307, \delta_e = 1.0657, \delta_a = 0, \delta_r = 0 \tag{27}$$

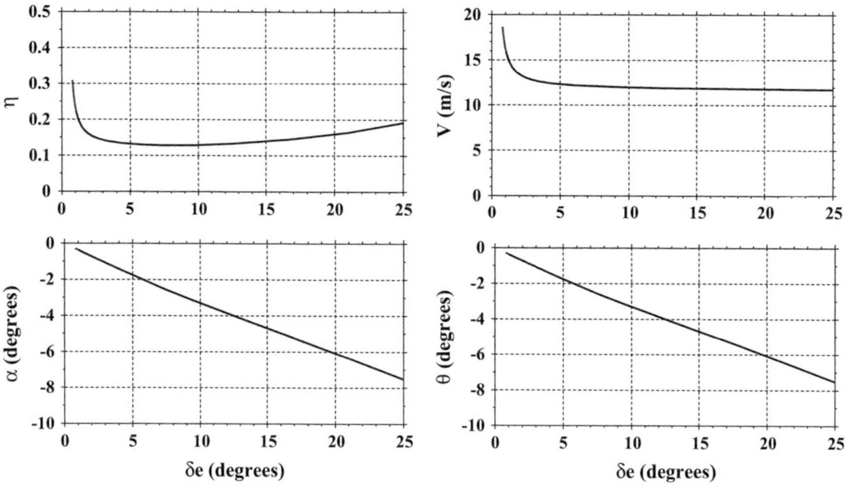

Fig. 3 Bifurcation plots for straight and level flight maneuver

Airship is an underactuated system with eight state variables and four control variables. Thus, choosing a stable trim as a maneuver design criterion helps in maintaining the integrity of the system even in the presence of parametric and nonparametric uncertainties. These set of trim solutions also help in selecting an optimal maneuver considering the complete dynamics of the system. For example, from the bifurcation plots, minimum energy straight and level flight maneuver corresponds to $\eta = 0.128$ and $\delta_e = 7.5°$, whereas minimum time straight and level flight maneuver corresponds to $\eta = 0.23$ and $\delta_e = 1.05°$. Thus, based on the applicability desired maneuver could be chosen from a set of stable trim solutions.

4.2 *Level Turn Maneuver*

Most of the studies on aircraft turn performance consider zero-sideslip bank-to-turn maneuvers. But in contrary, attainable bank angle computed from standard bifurcation analysis for maximum aileron deflection is only about $\pm 5°$, whereas skid angle for maximum rudder deflection is $\pm 30°$ which makes skid-to-turn an effective alternative for airships. The inherent limitation of roll in airship is also evident from [13] where roll angle is limited to $\pm 5°$ and [14] ignores roll from airship equations of motion. This is typically attributed to the dependency of airship on buoyancy for lifting as dynamic lift force is less. In addition, location of center of buoyancy is above the center of gravity which nullifies the moment about the roll axis [14]. This ascribes as an advantage on the functionality of airship, as payload for various applications is usually earth-pointing. Thus, this paper deals with level skid-to-turn maneuver design and evaluation for airship. On corollary, performance evaluation of turn maneuvers also results in an effective hover control strategy. Most applications of airship demand hovering at a stationary position coordinate. Airship is neutrally stable at hover condition, thereby making hover stabilization an important aspect of its control design.

The constraint equations for the evaluation of turn maneuvers are chosen prudently, such that the constraints are achievable by the available control capabilities. As sideslip angle corresponding to maximum rudder deflection is ±30°, sideslip angle of −25° with zero flight path and roll angle are considered for illustration. This choice of higher sideslip angle could help in characterizing the performance of airship by evaluating its maximum possible turn rate and minimum possible turn radius. Thus

$$\gamma = 0; \beta = -25°; \phi = 0 \tag{28}$$

As in straight and level flight maneuver, η, δ_r, and δ_a are freed to satisfy constraints on γ, β, and ϕ, respectively. As inferred from the bifurcation plots in Fig. 4, Hopf bifurcation occurs at the state and control trim vectors specified in Eqs. (29) and (30), respectively, with velocity in m/s and angles in degrees. But this trim corresponds to enforced throttle ratio of 1.2253, which an airship is incapable of achieving as the required thrust is more than the maximum available thrust. The onset of this instability is characterized by the pendulum mode becoming unstable at higher velocities.

$$\boldsymbol{x} = [16.5822, \; -1.6291, \; -25, 0.0320, 0, 1.1248, 0, \; -1.6291]^T \tag{29}$$

$$\eta = 1.2253, \delta_e = 5.4214, \delta_a = -3.3578, \delta_r = -19.5993 \tag{30}$$

In Fig. 5, turn rate (ω) is computed by solving rotational kinematic equation of yaw rate $(\dot{\psi})$ [Eq. (9)] and turn radius (R) is obtained by dividing velocity of the airship by turn rate, i.e., $R = V/\omega$. On thorough perusal of Figs. 4 and 5, it could be inferred that the maximum possible fastest turn and minimum possible tightest turn corresponds to elevator deflection of about 20°, with consideration on state and control limitations (δ_r reaches its saturation limit nearly after $\delta_e = 20°$). Thus, maximum turn rate and minimum turn radius are approximately 1°/s and 730 m, respectively. It is also noteworthy that the tightest turn demands comparatively a higher thrust.

4.3 *Ascend and Descend Maneuvers*

This section formulates feasible ascend and descend trajectories, emphasizing lateral excursions minimization with considerations on state and control constraints. The buoyant force experienced by the airship is given by

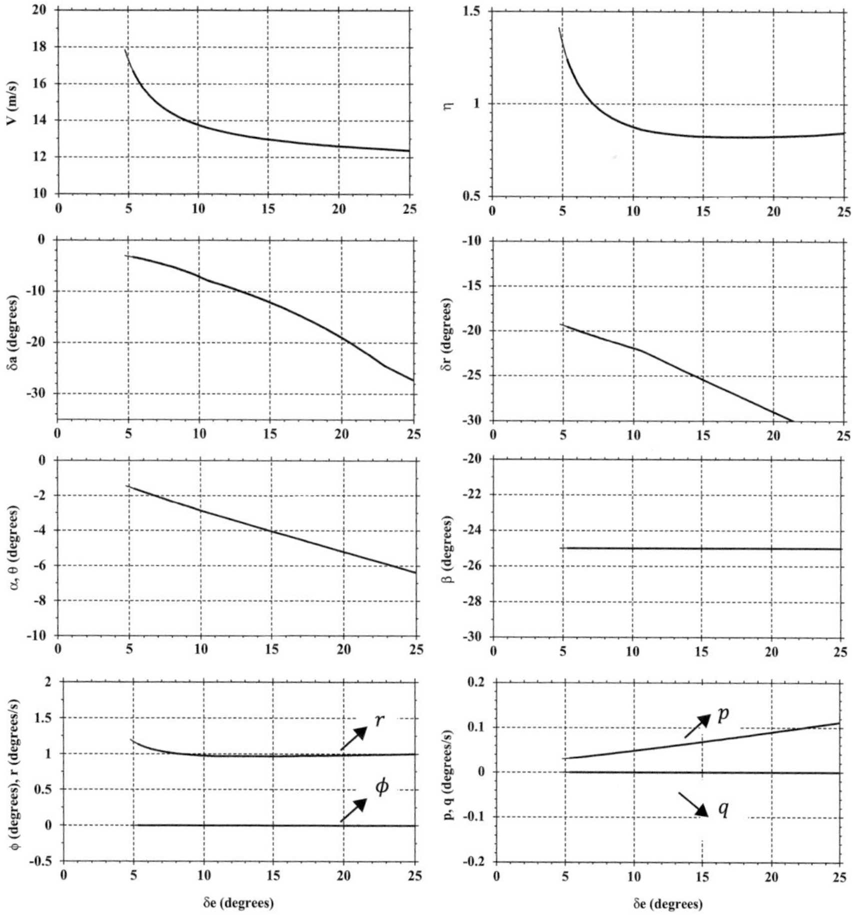

Fig. 4 Bifurcation plots for level turn maneuver

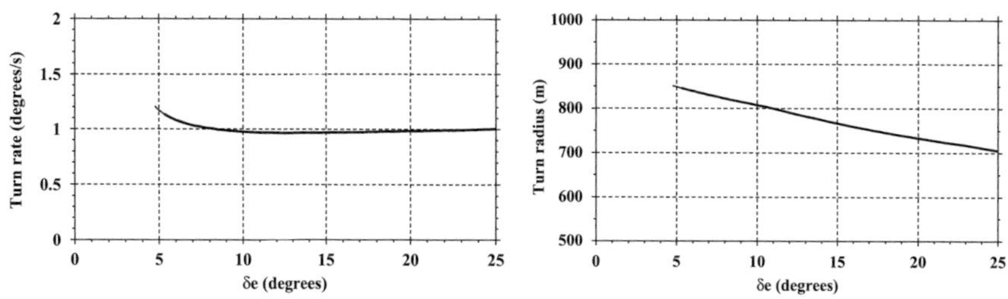

Fig. 5 Performance of level turn maneuver

$$B = U(h)\rho(h) \tag{31}$$

where U is the volume of the body and ρ is the mean density of air surrounding the body. The envelope of airship could handle only small differential pressure. Internal pressure is regulated using ballonet system to maintain zero static lift, i.e., $B = W$ throughout the flight regime. Airship is considered an open system with ballonets open to the atmosphere. Thus, change in ballonet's volume during ascend and descend changes the volume of the system with the mass of airship remaining constant.

Airship employing thrust vectoring exhibit enhanced ascend and descend performance. Thus, the vector angle for ascend is taken as 45°, based on exhaustive flight testing performed on the Airship Industries Skyship 500/600 series [8]. To achieve a minimum time ascend, maximum flight path angle and maximum available power, P_{m} is utilized. With the assumption that the propeller and motor efficiency is 1, maximum available thrust, T_{m} is given by

$$T_{\mathrm{m}} = P_{\mathrm{m}}/V \tag{32}$$

The constraints for minimum time ascend is given by

$$\gamma = 10°; \beta = 0; \phi = 0; \eta = 1; \zeta = 45° \tag{33}$$

During ascend, ρ has a nonlinear variation with respect to the body's altitude. This variation affects the dynamic pressure of airship which in turn affects the aerodynamic forces and moments. Thus, change in ρ is adapted by inducing it in control vector and using it as a continuation parameter. For the constraints specified in Eq. (33), δ_e, δ_a, and δ_r are freed. Since, constraints on lateral directional variables are zero, trim values of lateral directional controls, δ_a and δ_r are zero, thus, p and r are as well zero. Elevator control schedule with maximum available thrust and longitudinal state variables are plotted in Fig. 6. It could be inferred that the Hopf bifurcation occurs at the state and control trim vectors specified in Eqs. (34) and (35), respectively, with density in kg/m^3, velocity in m/s, and angles in degrees.

$$\boldsymbol{x} = [16.6446, 1.3431, 0, 0, 0, 0, 0, 11.3926]^T \tag{34}$$

$$\eta = 1, \zeta = 45, \delta_e = -12.3505, \delta_a = 0, \delta_r = 0, \rho = 0.1557 \tag{35}$$

This minimal time ascend maneuver demands a substantial lateral air space of around 100 km, which is difficult to ensure in the air traffic zone. The lateral excursion is largely reduced by imparting a skid angle, which results in a helical ascend trajectory with radius of helix corresponding to the induced skid angle. A larger skid angle reduces the radius of turn but it must be ensured that the proposed skid angle is achievable by the available control efforts. For helical ascend maneuver, constraints are chosen as

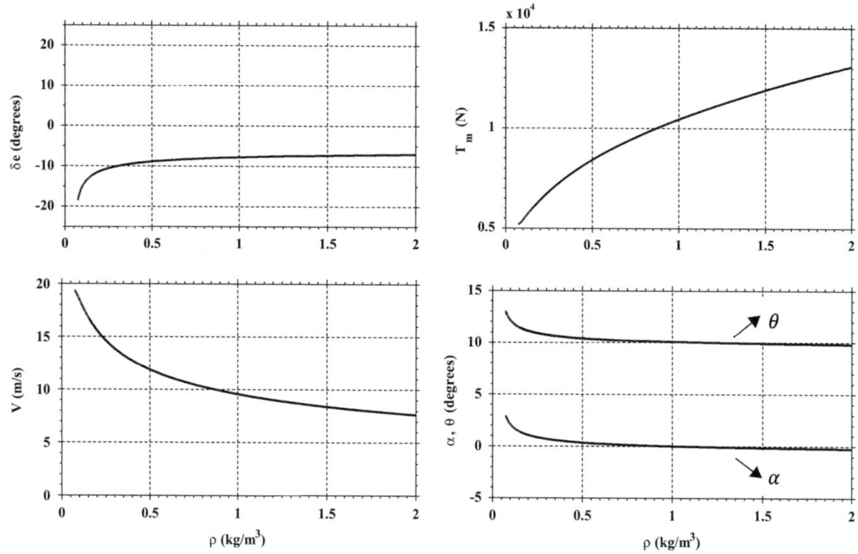

Fig. 6 Response for minimal time ascend

$$\gamma = 10^\circ; \beta = 5^\circ; \phi = 0; \eta = 1; \zeta = 45^\circ \tag{36}$$

Control schedules and the response of state variables for the considered helical ascend are plotted in Figs. 7 and 8. Hopf bifurcation occurs at the state and control trim vectors specified in Eqs. (37) and (38), respectively, with density in kg/m^3, velocity in m/s, and angles in degrees.

$$\boldsymbol{x} = [16.0369, 1.4108, 5, 0.0652, 0, -0.3203, 0, 11.4991]^T \tag{37}$$

$$\eta = 1, \zeta = 45, \delta_e = -13.7454, \delta_a = 0.4804, \delta_r = 7.0004, \rho = 0.1550 \tag{38}$$

Both minimal time and helical ascend maneuvers with $\zeta = 45^\circ$ and $\eta = 1$ become unstable at a density of around 0.15. This corresponds to an altitude of about 16,670 m, which is conveniently above the air traffic zone and other intrusions. Velocity also shoots up to its maximum allowable value. This demands a change in approach to ensure the stability of airship after the airship reaches an altitude of 16,670 m.

Minimal lateral excursion need not be considered a hard constraint above an altitude of 16,670 m, aiding zero skid angle consideration. Exerting maximum available thrust during ascend increases the velocity of airship as inferred from Figs. 6 and 8. This calls for curbing the velocity below its maximum value to ensure safety. Thus, a constraint in velocity of 13.5 m/s is considered with $\gamma = 3^\circ$. The tilt angle is brought back to zero. This results in the constraint equations given by

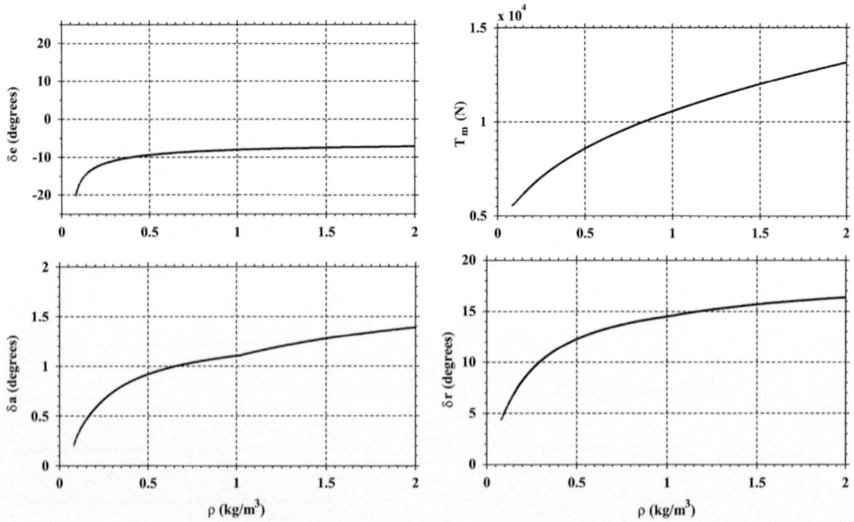

Fig. 7 Control schedule for helical ascend

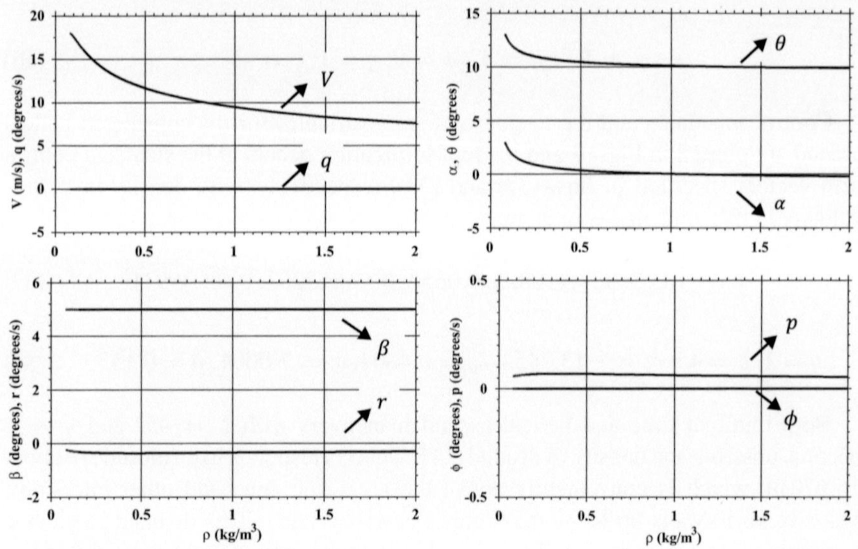

Fig. 8 Response of state variables for helical ascend

$$V = 13.5\,\text{m/s};\, \gamma = 3^{\circ};\, \beta = 0;\, \phi = 0;\, \zeta = 0 \tag{39}$$

η and δ_e are freed to achieve constraints on V and γ with the maximum available thrust taken to be 6000 N. It is also evident that δ_a and δ_r are zero to achieve zero skid and roll angle. Response of non-zero state and control variables with respect to

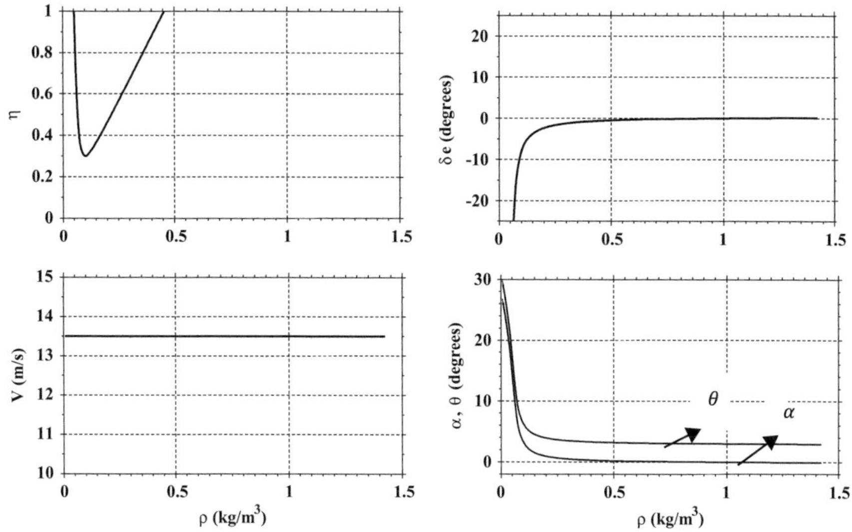

Fig. 9 Response for longitudinal only maneuver with constraints on V and γ

the specified set of constraints is depicted in Fig. 9. From the plots, it is inferred that the flight path angle is maintained at 3° and the commanded velocity is unachievable for densities above 0.45 kg/m^3, as throttle ratio shoots above 1. Thus, for velocity constraint maneuver during ascend, velocity should be judiciously chosen through thorough perusal on velocity profile corresponding to maximum available thrust in Fig. 6. This maneuver could be used in the final neck of ascend, beyond the air traffic prone zone of the lower stratosphere.

The same approach could be carried out for descend maneuvers design. While descending, constrained longitudinal only maneuver with $V = 13.5$ m/s and $\gamma = -3°$ is used till an altitude of 16 km. After which, a helical descend maneuver with $\gamma = -10°$ is employed. A narrow regime of stability in ascend and descend maneuver design emphasizes the importance of the proposed maneuver design approach. A randomly chosen maneuver might not work under the state and control constraints of airship. This explains scant attention on ascend and descend in the literature pertaining to the control of stratospheric airship.

5 Validation

This section illustrates the feasibility of proposed maneuvers with control schedules in open-loop. It also helps in analyzing the performance of computational bifurcation methodology in maneuver design and optimization approaches.

5.1 Level Performance Analysis

A scenario of executing the tightest possible turn from a steady level flight condition is considered. A feasible state and control trim corresponding to a stable straight and level condition is chosen from Fig. 3 and is tabulated in Eqs. (40) and (41), respectively, with velocity in m/s and angles in degrees.

$$\boldsymbol{x} = [14.6671,\ -0.4810, 0, 0, 0, 0, 0,\ -0.4810]^T \tag{40}$$

$$\eta = 0.1913, \delta_e = 1.3145, \delta_a = 0, \delta_r = 0 \tag{41}$$

Airship is simulated with controls in Eq. (41) for 500 s. After which it is commanded to execute the minimum radius turn maneuver. From Fig. 4, state and control trim corresponding to a stable and achievable minimum radius of turn is chosen and is given in Eqs. (42) and (43), respectively, with velocity in m/s and angles in degrees.

$$\boldsymbol{x} = [12.7204,\ -4.7627,\ -25, 0.0812, 0, 0.9750, 0,\ -4.7627]^T \tag{42}$$

$$\eta = 0.8216, \delta_e = 18.0558, \delta_a = -15.9716, \delta_r = -27.6369 \tag{43}$$

Response of airship during the execution of proposed maneuver is plotted in Figs. 10 and 11. For the initial 500 s, airship travels with the velocity of about 14.6 m/s covering a distance of 7300 m. Airship is then commanded to execute level turn maneuver with minimum possible turn radius. It is inferred from Fig. 10 that the radius of turn is approximately 740 m, which matches with the turn radius computed and plotted in Fig. 5. In order to maintain zero flight path angle, η is freed as discussed in Sect. 4. But controlling γ demands controlling both α and θ, i.e., two state variables are controlled using a single control input. Such issues of underactuation are effectively handled by this maneuver design approach.

This illustration also helps in analyzing the hover capabilities of airship. Hover of airship at the pressure altitude is a neutrally stable flight condition [15]. Thus,

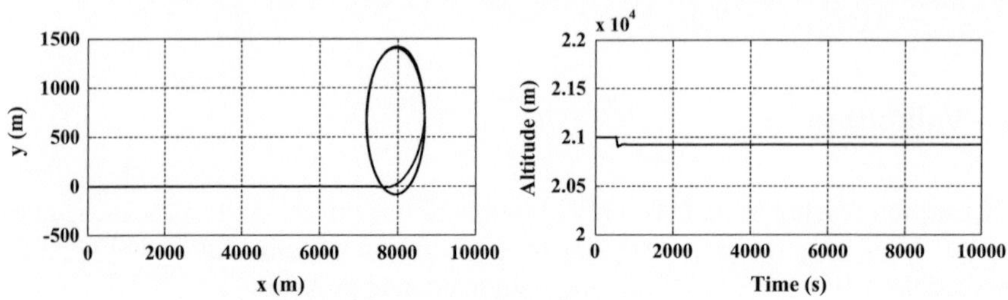

Fig. 10 Position trajectories for the considered maneuver

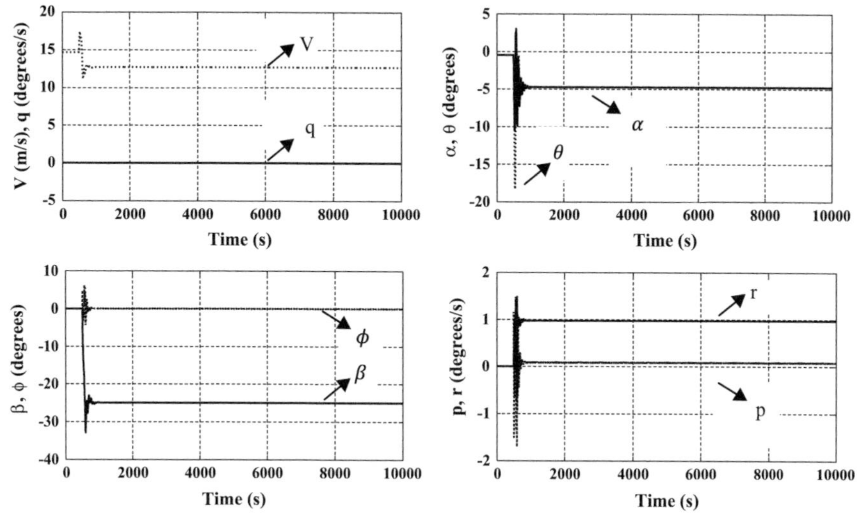

Fig. 11 Response of state variables for the considered maneuvers

maintaining its position coordinates amidst strong gust becomes a challenging task in autonomous airship maneuvering. Few works demonstrate the challenges faced by airship during hover in turbulent atmospheric environments and infer that significant thrust and control power are required to station-keep [16, 17], with the risk on system's integrity at worst case scenarios. In such situations, the proposed tightest turn maneuver could be commanded with the backing of robust controller to achieve it. This causes airship to suffer only a minor change from the station-kept position coordinates. If the thrust demand shoots up further, skid angle could be considerably reduced and maneuver is appropriately redesigned while ensuring the safety and integrity of the airship.

5.2 *Ascend Maneuver Evaluation*

The feasibility of proposed ascend maneuvers with control schedules in open-loop is evaluated. The proposed helical and longitudinal only ascend maneuvers are evaluated using the simulation framework portrayed in Fig. 12. Based on the altitude of airship ($h = -z_E$), ρ is calculated using curve fitting of atmospheric density and altitude data. From sea level corresponding to $\rho = 1.2256\,\text{kg/m}^3$, through $h = 16{,}670\,\text{m}$ corresponding to $\rho = 0.15$, control schedules of helical ascend with respect to their corresponding densities (plotted in Fig. 7) are fed to airship model with $\eta = 1$ and $\zeta = 45°$. Subsequently, control schedules corresponding to longitudinal only ascend (plotted in Fig. 9) with $V = 13.5\,\text{m/s}$ and $\gamma = 3°$ are fed, with $T_{\text{m}} = 6000\,\text{N}$ and $\delta_a, \delta_r, \zeta = 0$, till the airship reaches its operational altitude

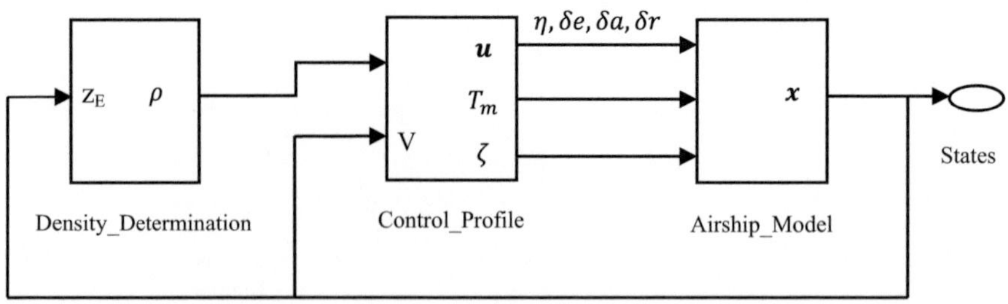

Fig. 12 Simulation framework for ascend maneuver validation

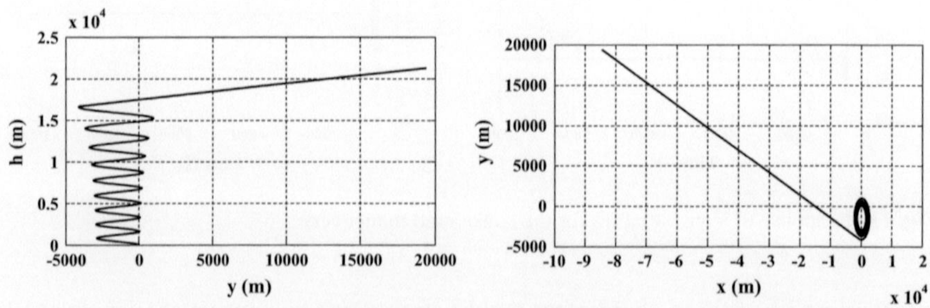

Fig. 13 Position trajectories for the proposed ascend maneuver

corresponding to density of around 0.075 kg/m^3. The response of state variables $\boldsymbol{x} = [V, \alpha, \beta, p, q, r, \phi, \theta, \psi, x_E, y_E, z_E]^T$ is recorded and fed back to determine corresponding density and control profile.

Position trajectories and state response of the implemented ascend maneuver are plotted in Figs. 13 and 14. It is evident that the helical ascend with $\gamma = 10°$ is carried out till an altitude of 16,500 m. During this phase, β is maintained at 5° to limit lateral excursion to a maximum of 5000 m, as inferred from Figs. 13 and 14. This is beneficial while climbing through air traffic and tropospheric interferences. Beyond this altitude, longitudinal only maneuver with constraints on V and γ is executed. The feasibility of the proposed ascend maneuver is thus successfully validated.

6 Conclusions

Autonomous maneuvering of aerial vehicles has gained predominant interest in the recent past. In that regard, one of the prime realms that receive scant emphasis is maneuver design. This work establishes an effective maneuver design technique using bifurcation-based continuation approach. This methodology takes into consideration the dynamics of the system with state and control constraints and generates a series of feasible solutions. A few major challenges in airship autonomy like stability

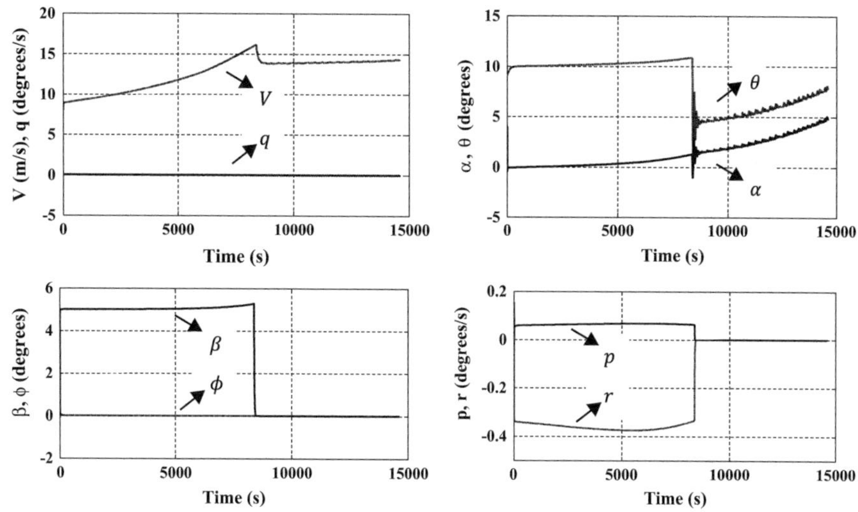

Fig. 14 Response of state variables for the proposed ascend maneuvers

during hover, ascend, and descend phases are addressed. The proposed maneuvers are then validated using an open-loop formulation with computed control profiles from bifurcation analysis. The flexibility to choose from a family of feasible solutions with different constraint sets makes this a unique platform for maneuver design. The reliability of controllers in uncertain environments could be greatly enhanced with the knowledge of feasible solutions, especially for underactuated systems.

References

1. Tsiotras P, Mesbahi M (2017) Toward an algorithmic control theory. J Guid Control Dyn 40(2):194–196
2. Lu P (2017) Introducing computational guidance and control. J Guid Control Dyn 40(2):193
3. Hima S, Bestaoui Y (2002) Motion generation on trim trajectories for an autonomous underatuated airship. In: Proceedings of the 4th international airship convention and exhibition, airship association, Cambridge, England
4. Repoulias F, Papadopoulos E (2007) Dynamically feasible trajectory and open-loop control design for unmanned airships. In: Proceedings of the 8th IFAC symposium on robot control, Vol 39, no 15, pp 592–597
5. Khatri AK, Singh J, Sinha NK (2012) Aircraft maneuver design using bifurcation analysis and sliding mode control techniques. J Guid Control Dyn 35(5):1435–1449
6. Dhooge A, Govaerts W, Kuznetsov AY (2003) MATCONT: A MATLAB package for numerical bifurcation analysis of ODEs. ACM Trans Math Softw 29(2):141–164
7. Vora AS, Sinha NK (2017) Direct methodology for constrained system analysis with applications to aircraft dynamics. J Aircr 54(6):2378–2385
8. Khoury GA (2012) Airship technology. Cambridge University Press, New York, USA

9. Rana V, Ajith K, Sinha NK, Amiatbha P, Sati SC (2015) Configuration analysis of stratospheric airship. In: Symposium on applied aerodynamics and design of aerospace vehicles, VSSC, Thiruvananthapuram, India
10. Sinha NK, Ananthkrishnan N (2014) Elementary flight dynamics with an introduction to bifurcation and continuation methods. CRC Press, Boca Raton, USA
11. Ananthkrishnan N, Sinha NK (2001) Level flight trim and stability analysis using extended bifurcation and continuation method. J Guid Control Dyn 24(6):1225–1228
12. Khatri AK, Singh J, Sinha NK (2013) Accessible regions for controlled aircraft maneuvering. J Guid Control Dyn 36(6):1829–1834
13. Lee S, Bang H (2007) Three-dimensional ascent trajectory optimization for stratospheric airship platforms in the jet stream. J Guid Control Dyn 30(5):1341–1351
14. Mueller JB, Zhao YJ, Garrard WL (2009) Optimal ascent trajectories for stratospheric airships using wind energy. J Guid Control Dyn 32(4):1232–1245
15. Cook MV, Lipscombe JM, Goineau F (2000) Analysis of the stability modes of the non-rigid airship. Aeronaut J 104(1036):279–290
16. Schmidt DK (2007) Modeling and near-space stationkeeping control of a large high-altitude airship. J Guid Control Dyn 30(2):540–547
17. Schmidt DK, Stevens J, Roney J (2007) Near-space station-keeping performance of a large high-altitude notional airship. J Aircr 44(2):611–615

Estimation of Stability Derivatives Due to Translational Motion of Various LTA Vehicles Using CFD

Anoop Sasidharan, Ratna Kishore Velamati, Sheeja Janardhanan, Venkata Ramana Murthy Oruganti, and Akram Mohammad

1 Introduction

Wind data is an inevitable part of the wind energy sector. The major applications of wind data are the dispatching of load [1], forecasting demand [2], planning the generation [3] and maintenance of the mechanical structures [4]. The conventional method of wind data collection in the wind energy harvesting sites is based on the pole based methods. This method has a few drawbacks, such as the need for a permanent structural setup which may cause environmental and aesthetic impacts on the site, the inability to vary the altitude of the structure, a long gestation period and cannot be relocated or reused easily [5]. A tethered aerostat can be considered as an alternate method for low-altitude wind measurement. Compared to the conventional method, the proposed method has the following advantages; it gives freedom to vary the altitude of operation, it does not require any permanent structure for the deployment, takes less time for deploying at a site and can be relocated easily to different locations and reused. The feasibility of such a system should be investigated with the help of mathematical models.

Mathematical modelling of complex systems such as lighter-than-air (LTA) vehicles plays a vital role in understanding and predicting the unsteady aerodynamic behaviour of that system [6]. Since the functional form of aerodynamic effects [7] in the form of forces and moments on the aerial vehicle is not readily available for

A. Sasidharan · R. K. Velamati (✉) · V. R. M. Oruganti
Department of Electrical and Electronics Engineering, Amrita School of Engineering, Coimbatore, Amrita Vishwa Vidyapeetham, Coimbatore, India
e-mail: v_ratnakishore@cb.amrita.edu

S. Janardhanan
School of Naval Architecture and Ocean Engineering, Indian Maritime University, Visakhapatnam, India

A. Mohammad
Department of Aerospace Engineering, King Abdulaziz University, Jeddah, Saudi Arabia

D. Shukla (ed.), *Lighter Than Air Systems*, Lecture Notes in Mechanical Engineering,
https://doi.org/10.1007/978-981-19-6049-9_17

direct measurement, mathematical derivatives are needed to model these forces and moments [8, 9]. Derivatives are a measure of the degree of variation of a function or a parameter with respect to input or another parameter. In the field of dynamics, those derivatives that can represent the model's sensitivity to some motion variables are termed stability derivatives because these derivatives help simplify the model so that the stability analysis and control of the model become easy. Aerodynamic models using stability derivatives are considered an accurate approach to the mathematical modelling of aerial vehicles.

The estimation or extraction of the stability derivatives was reported mainly using flight tests, wind tunnel tests and numerical methods. Some of the essential works reported in these classes are presented here. The estimation of the aerodynamic derivatives for the LTA vehicles using a semi-empirical procedure was reported in [10]. They have considered the hull-fin interference using an analytical model and the static wind tunnel data by which the dynamic motion of the vehicle can be represented well. As an extension, they have investigated the longitudinal and lateral stability derivatives for the TCOM 71M aerostat [11]. The model also verified the static and dynamic stability of the aerostat obtained using the stability derivatives. A neural network-based prediction of the dynamic derivatives of an aircraft at high angles of attack was presented by [12]. They provided decent results for pitching moment derivatives compared to the wind tunnel experimental data. A wind tunnel-based stability derivative estimation for a hybrid buoyant aerial vehicle is presented by [13]. They compared the stability of the vehicle with wings and without wings. Even though the physical realism of the experimental methods such as flight tests and wind tunnel tests is highly appreciable, they have some limitations such as scaling errors, blocking effects and high cost. A system identification approach for obtaining the stability derivatives of an LTA vehicle undergoing swing oscillation was presented in [14]. Computational methods were also reported to estimate the stability derivatives of LTA vehicles [15].

Considering the fact that the computational fluid dynamics (CFD) solvers are now capable of handling high-performance tasks due to the developments in digital computing, their application in the extraction of stability derivatives is being largely explored by the research community. Most of the limitations of the experimental method, such as the wind tunnel tests, can be overcome by the CFD based methods. A forced oscillation of the body was used for extracting the stability derivatives. The motion variables other than the oscillating parameter are considered zero as per the linear theory. A forced periodic oscillation is imposed on the respective degree of freedom of the model to calculate the respective dynamic derivative. For example, the moment stability derivative due to the pitch rate can be extracted by imposing a periodic angular oscillation about the lateral axis and centre of the moment [9]. The measured moment response will be used to extract the derivative based on the Fourier series approximation. Such a methodology for estimating stability derivatives using CFD for the Zhiyuan airship is presented in [16]. Stability derivatives involved in the pitch and heave motions of the airship were investigated using forced sinusoidal oscillations. They have presented the effect of the tail fins on stability as well. Stability derivatives prediction for a full aircraft configuration was investigated to represent the unsteady aerodynamic load on the flight dynamics for transonic speeds and larger angles of attack as in [17, 18]. An investigation of the stability of hybrid airships using the analysis of the stability derivatives was presented by [19]. The added mass and inertia terms extraction for an underwater vehicle were reported in [20]. They have compared their results with the results from an experimental procedure and theoretical results. Using stability derivatives addressed by [21], the aerodynamic shape optimisation considered the derivatives as constraints in the optimisation problem.

The forced sinusoidal oscillation-based stability derivative extraction is the most popular method due to the well developed Fourier theory and the easy practical realisation [22, 23]. There are also works reported with different methodologies, such as the work reported in [24]. The history effect and the actuation of the body using piece-wise velocity functions were the basis of their methodology. Even though the rotational dynamics were not able to capture, the methodology was validated with existing data. A detailed literature review of the existing methodologies for stability derivative extraction using CFD can be found in [25].

In this paper, the authors present the stability derivative estimation of aerostat using CFD-based methodology. Stability derivatives of the aerostat due to translational oscillations are extracted by applying a translational forced sinusoidal oscillation along with the vertical and axial directions. A steady wind condition is considered for the study. As the geometry of the aerostat has a significant influence on the aerodynamic stability, the effect of the aerostat envelope shapes in the context of stability derivatives which is investigated as well. It will also help to understand the methodology's efficacy and generalisation thoroughly. There are many aerostat/airship geometries available in the open literature. The NPL shape has been designed and developed by the National Physical Laboratory for reducing the chances of flow separation [26]. It was achieved by providing a continuous variation in the radius of curvature of the profile. The GNVR shape has been reported to be a low drag profile [26]. Both of these shapes represented a combination of different simple geometries, such as ellipse and parabola. It made them feasible for geometry optimisation. A high-altitude application was reported, which was claimed to have a minimum weight to increase the payload and stability (HAA shape) [27]. All the above-mentioned shapes were proposed for stratospheric operation. A bionic design for the airship has been reported to minimise drag (bionic shape) [28]. It was achieved by imitating the morphological features of an aquatic animal (*Physalia physalis*). Other low drag profiles were also reported for specific applications, such as the Zhiyuan shape [16] and Wang shape [29].

The rest of the paper is structured as follows; Sect. 2 presents the methodology involved in the stability derivative extraction of aerostat using CFD analysis and the various aerostat models used for the analysis. Section 4 presents the numerical simulation setup considered for the current study. Section 5 presents the results obtained from the analysis of the stability derivatives due to the heave and surge oscillations and a comparison of the derivatives among the four aerostats. Finally, Sect. 6 concludes the work with the major findings and the future scope of the study.

2 Stability Derivatives Extraction Methodology

In this work, the longitudinal motion of the aerostat is considered. The lateral motion of the aerostat is assumed to be negated by the control surfaces. The motion variables involved in the decoupled linear longitudinal dynamic model are u, w and q. The longitudinal model of the aerostat can be expressed as [30],

$$
\begin{aligned}
&m_x\dot{u} + (ma_z - \dot{D}_{\dot{q}})\dot{q} = D_a + D_T + D_g \\
&m_z\dot{w} - (ma_x + \dot{L}_{\dot{q}})\dot{q} = L_a + L_T + L_g \\
&J_y\dot{q} + (ma_z - \dot{M}_{\dot{u}})\dot{u} - (ma_x + \dot{M}_{\dot{w}})\dot{w} = M_a + M_T + M_g
\end{aligned}
\tag{1}
$$

where, D_a, L_a and M_a are the aerodynamic forces and moment, D_T, L_T and M_T are the forces and moment due to tethers and D_g, L_g and M_g are the forces and moment due to buoyancy and gravity.

In the current study, the tether dynamics and the forces due to gravity are not included, and for simplicity, they will be named as external forces (F_{ext}) in the rest of the paper. The aerodynamic part of the longitudinal dynamic model of the aerostat can be represented as follows,

$$
\begin{aligned}
D_a &= D_e + \dot{D}_u u + \dot{D}_w w + (\dot{D}_q - m_z W_e)q \\
L_a &= L_e + \dot{L}_u u + \dot{L}_w w + (\dot{L}_q + m_x U_e)q \\
M_a &= M_e + \dot{M}_u u + \dot{M}_w w + (\dot{M}_q - ma_x U_e - ma_z W_e)q
\end{aligned}
\tag{2}
$$

where W_e and U_e are the component of steady velocity along the vertical and axial direction, D_e, L_e and M_e are the static drag, lift and moment derivatives, $\dot{D}_u$, $\dot{L}_u$, $\dot{M}_u$ are the drag, lift, and moment derivatives due to axial velocity, $\dot{D}_w$, $\dot{L}_w$, $\dot{M}_w$ are the drag, lift, and moment derivatives due to vertical velocity, $\dot{D}_q$, $\dot{L}_q$ and $\dot{M}_q$ are the drag, lift and moment derivatives due to pitch.

Thus, Eq. (1) can be expressed in matrix form as given in Eq. (3). The terms with dots are the stability derivatives involved in the longitudinal dynamics of the aerostat.

$$
\begin{aligned}
&\begin{bmatrix} m - \dot{D}_{\dot{u}} & 0 & ma_z - \dot{D}_{\dot{q}} \\ 0 & m - \dot{L}_{\dot{w}} & ma_x + \dot{L}_{\dot{q}} \\ ma_z - \dot{M}_{\dot{u}} & ma_x + \dot{M}_{\dot{w}} & I_y - \dot{M}_{\dot{q}} \end{bmatrix} \begin{bmatrix} \dot{u} \\ \dot{w} \\ \dot{q} \end{bmatrix} = \\
&\begin{bmatrix} D_e \\ L_e \\ M_e \end{bmatrix} + \begin{bmatrix} \dot{D}_u & \dot{D}_w & \dot{D}_q - m_z W_e \\ \dot{L}_u & \dot{L}_w & \dot{L}_q + m_x U_e \\ \dot{M}_u & \dot{M}_w & \dot{M}_q - ma_x U_e - ma_z W_e \end{bmatrix} \begin{bmatrix} u \\ w \\ q \end{bmatrix} + F_{\text{ext}}
\end{aligned}
\tag{3}
$$

The model variables included in the longitudinal model of the aerostat are u, w, q, D, L and M. These longitudinal dynamic variables are predominant in the heave, surge and pitch motions of the aerostat. Figure 1a shows the orientation of the aerostat axis, wind direction and the motions considered in this study.

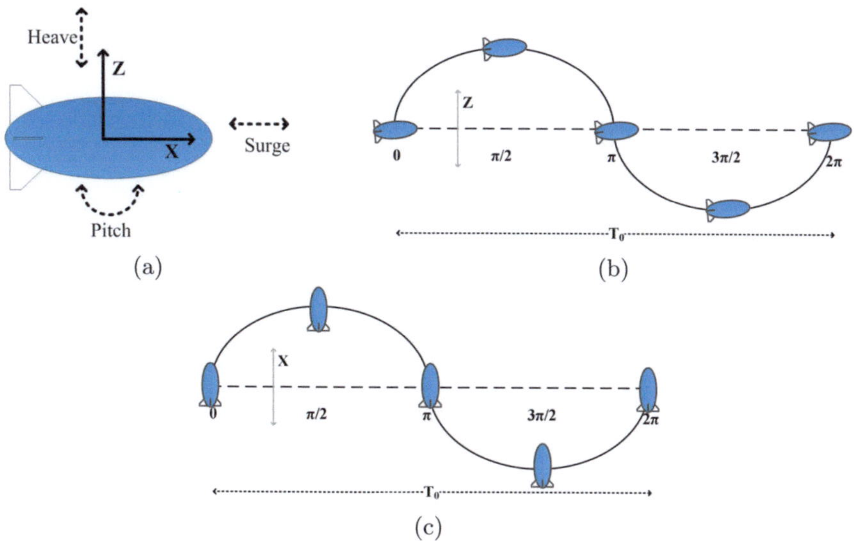

Fig. 1 **a** Body fixed reference axis and the directions of motions considered, **b** heave motion and **c** surge motion

This paper investigates the stability derivatives due to the surge and heave oscillations of the aerostat. The methodology used for the extraction of the stability derivatives includes the simulation of the model for steady-state and then transient and finally the oscillations in the respective directions. The forces and moment (D, L and M) are measured for the oscillated motions. A stable cycle of data will be used from these measured responses to extract the stability derivatives using the Fourier series-based method discussed as follows.

The stability derivatives associated with the translational forward motion (along the X-axis) can be expressed as the linear decomposed forces and moment as follows,

$$\begin{aligned} D_a^s &= D_e + \dot{D}_u u + \dot{D}_{\dot{u}} \dot{u} \\ L_a^s &= L_e + \dot{L}_u u + \dot{L}_{\dot{u}} \dot{u} \\ M_a^s &= M_e + \dot{M}_u u + \dot{M}_{\dot{u}} \dot{u} \end{aligned} \tag{4}$$

A forced sinusoidal oscillatory motion will be induced in the aerostat axial direction, as shown in Fig. 1b. The motion can be defined by the displacement $x = u_0 \sin \omega_s t$, velocity $u = \omega_s u_0 \cos \omega_s t$ and acceleration $\dot{u} = -\omega_s^2 u_0 \sin \omega_s t$. Substituting the values of u and $\dot{u}$ into Eq. (4) will result in,

$$\begin{aligned} D_a^s &= D_e + \dot{D}_u \omega_s u_0 \cos \omega_s t - \dot{D}_{\dot{u}} \omega_s^2 u_0 \sin \omega_s t \\ L_a^s &= L_e + \dot{L}_u \omega_s u_0 \cos \omega_s t - \dot{L}_{\dot{u}} \omega_s^2 u_0 \sin \omega_s t \\ M_a^s &= M_e + \dot{M}_u \omega_s u_0 \cos \omega_s t - \dot{M}_{\dot{u}} \omega_s^2 u_0 \sin \omega_s t \end{aligned} \tag{5}$$

As Eq. (5) comprises the sine and cosine components, the coefficients can be compared with the corresponding Fourier series representations of the forces and moment with higher-order terms neglected as given in Eq. (6).

$$\begin{aligned} D_a &= a_0/2 + a_1 \cos\omega_s t + a_2 \sin\omega_s t \\ L_a &= b_0/2 + b_1 \cos\omega_s t + b_2 \sin\omega_s t \\ M_a &= c_0/2 + c_1 \cos\omega_s t + c_2 \sin\omega_s t \end{aligned} \tag{6}$$

By comparing the coefficients of in phase and out of phase components of Eq. (6) and Eq. (5),

$$\begin{aligned} \dot{D}_u &= a_1/(\omega_s u_0); & \dot{L}_u &= b_1/(\omega_s u_0) \\ \dot{M}_u &= c_1/(\omega_s u_0); & \dot{D}_{\dot{u}} &= -a_2/(\omega_s^2 u_0) \\ \dot{L}_{\dot{u}} &= -b_2/(\omega_s^2 u_0); & \dot{M}_{\dot{u}} &= -c_2/(\omega_s^2 u_0) \end{aligned} \tag{7}$$

where,

$$a_1 = \frac{2}{T_0}\int_{-T_0/2}^{T_0/2} D_a(t)\cos(\omega_s t)\mathrm{d}t \qquad a_2 = \frac{2}{T_0}\int_{-T_0/2}^{T_0/2} D_a(t)\sin(\omega_s t)\mathrm{d}t$$

$$b_1 = \frac{2}{T_0}\int_{-T_0/2}^{T_0/2} L_a(t)\cos(\omega_s t)\mathrm{d}t \qquad b_2 = \frac{2}{T_0}\int_{-T_0/2}^{T_0/2} L_a(t)\sin(\omega_s t)\mathrm{d}t$$

$$c_1 = \frac{2}{T_0}\int_{-T_0/2}^{T_0/2} M_a(t)\cos(\omega_s t)\mathrm{d}t \qquad c_2 = \frac{2}{T_0}\int_{-T_0/2}^{T_0/2} M_a(t)\sin(\omega_s t)\mathrm{d}t$$

According to Eq. (7), a full cycle response of drag, lift and moment for the forced oscillation along the axial direction will give the six stability derivatives due to axial velocity and acceleration. In a similar manner, the stability derivatives due to heave oscillations can be estimated.

3 Aerostat Shapes

In this study, a four-winged aerostat with a '+' orientation of the wings is considered. The length of the aerostat is fixed as 13.5 m by considering the required volume for the desired payload weight. Four different aerostat envelope shapes obtained from the open literature were considered for the comparison of stability derivative results obtained from the current study. They are: the Zhiyuan shape [16], GNVR shape [26], NPL shape [26] and HAA shape [27], as shown in Fig. 2. These shapes were selected by considering the variation of their leading edge geometry, diameter and the

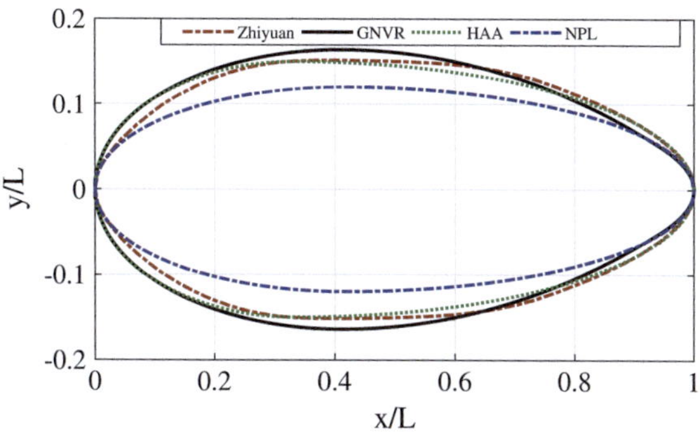

Fig. 2 Aerostat shapes considered for the current study

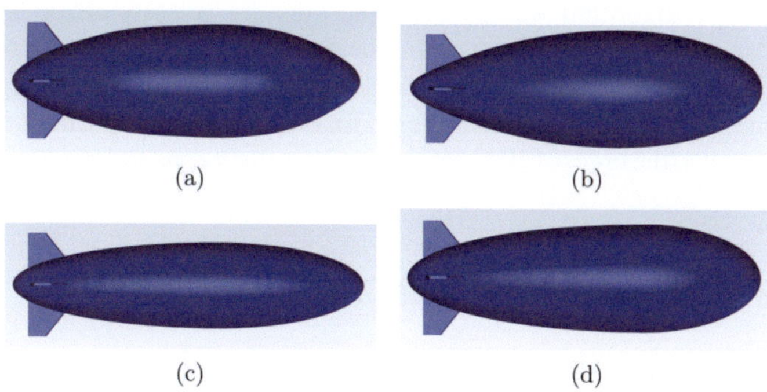

Fig. 3 Aerostat 3D models; **a** Zhiyuan, **b** GNVR, **c** NPL and **d** HAA shape

lift-drag characteristics. The 3D models used for the simulation analysis are shown in Fig. 3. The maximum diameter is 4.09 m for the Zhiyuan shape, 4.55 m for the GNVR shape, 3.33 m for the NPL shape and 4.09 m for the HAA shape. Among the four aerostat shapes considered, Zhiyuan has the maximum volume, and NPL has the least volume.

4 Numerical Simulation

4.1 Governing Equations

The current study involves the simulations done using a commercial CFD solver FLUENT 20.0. The 3D, incompressible, unsteady flow was solved using the Reynolds Averaged Navier—Stokes (RANS) equations [31]. The governing equations can be expressed as,

$$\nabla \cdot \vec{\mathbf{V}} = 0$$
$$\rho \frac{\partial \vec{V}}{\partial t} + \rho \nabla \cdot \left(\vec{\mathbf{V}}\, \vec{\mathbf{V}} \right) = -\nabla p + \nabla \cdot \left(\bar{\bar{\tau}} \right) + \rho \vec{g} \tag{8}$$

where ρ is the density, $\vec{\mathbf{V}}$ is the velocity, p is the pressure and $\vec{g}$ is the acceleration due to gravity. The stress tensor $\bar{\bar{\tau}}$ can be expressed as,

$$\bar{\bar{\tau}} = (\mu + \mu_t) \left[\nabla \vec{V} + \left(\vec{V} \right)^T - \frac{2}{3} \nabla \cdot \vec{V} \mathbf{I} \right]$$

where μ is the dynamic viscosity, $\boldsymbol{\mu}_t$ is the turbulence viscosity and $\mathbf{I}$ is an identity matrix.

The SIMPLE algorithm was used for the pressure velocity coupling, and the second-order upwind scheme was used for the spatial discretization of the convective terms. A standard interpolation scheme was used for the pressure interpolation, and an iterative time advancement scheme was used for time advancement. The convergence criterion was set to the order of 10^{-5} for continuity, velocity and turbulence quantities.

4.2 Computational Domain and Grid

The computational domain used for the simulation is shown in Fig. 4. The dimensions for the domain are selected after an initial domain independence study. The outermost box volume shown in Fig. 4 has a velocity inlet on four faces, a pressure outlet on one face and symmetry on the other. The outer and intermediate volumes were common for all the aerostat simulations presented in this paper. Dimensions of the domain are specified as a function of the aerostat model length. The dynamic mesh feature

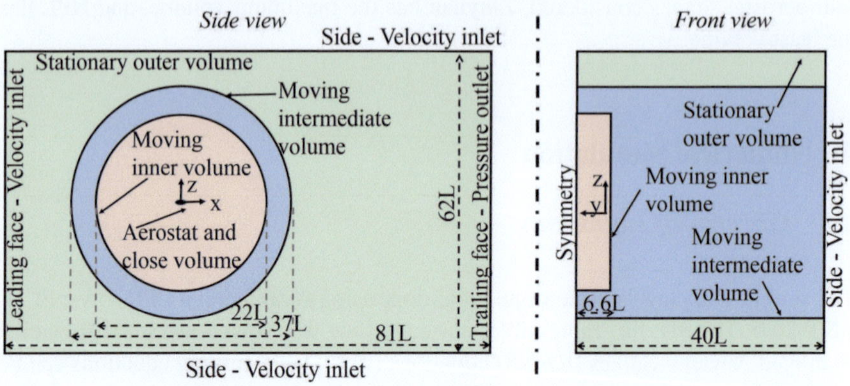

Fig. 4 Computational domain for the current study with the boundary conditions

realised the heave and surge motion simulations. The inner and intermediate volumes were specified as rigid bodies, and the outer volume was kept stationary for the surge and heave motions.

The computational grid used for the current study was made using GAMBIT and FLUENT meshing. The grid for outer and intermediate volumes (refer Fig. 4) was done using GAMBIT with a total of 0.1 M tetrahedral-hexahedral hybrid cells. Comparatively, the coarse grid was used for the outer volumes. Grid for the inner volume with aerostat (different for each aerostat shape) was done using ANSYS FLUENT meshing, as shown in Fig. 5. The range of the number of cells used for each shape is 3–3.5 M. Cut cell mesh topology was used for the inner volumes. The proximity of the aerostat was meshed fine using a refining body of influence in the shape of an ellipse. The inner volume mesh has 50 $\times$ 50 $\times$ 15 cells along the X, Y and Z Cartesian axes. The viscous boundary layer was specified with $5E-005$ as the first layer thickness, 1.35 as the stretching ratio and had 18 layers for the aerostat surface. With the above-specified grid configuration, the average value of the wall '$Y+$' is found to be 0.435 with a minimum of 0.0349 and a maximum of 1.

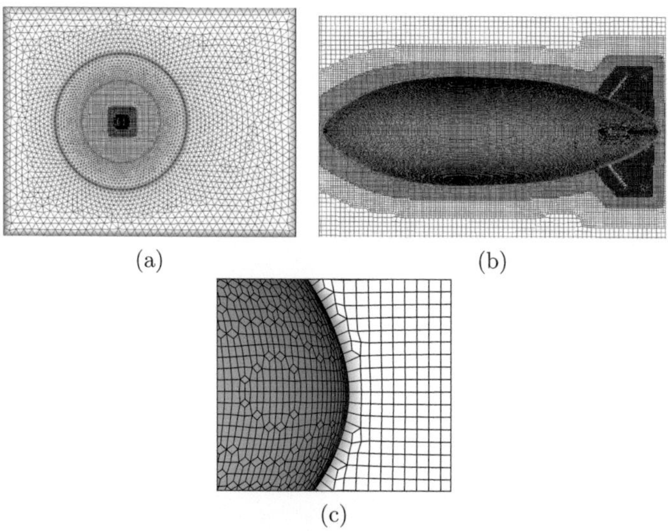

Fig. 5 Computational grid; **a** full domain, **b** close to the surface and **c** leading edge section with boundary layers

5 Results and Discussion

5.1 Grid and Time Step Independence Study

The efficacy of the mesh used for the analysis was tested for different grid refinements, as shown in Fig. 6a, and the simulation time step independence study was performed, as shown in Fig. 6b. The simulation was done by providing the aerostat with an unsteady velocity in the form of a sinusoidal gust. The coefficient of drag and moment was measured respectively for the comparison. The results show that the grid used for the study is independent of the grid refinement and the time step.

5.2 Validation of the Grid Motion

The dynamic mesh method for the heave and surge oscillations was validated using the AGARD CT1 [32]. NACA 0012 airfoil was used for the validation. A comparison of the lift coefficient and the pitching moment coefficient with the experimental data is shown in Fig. 7. An appreciable similarity is there with the results, thus validating the grid motion feasibility for further analysis.

5.3 Validation of the Stability Derivative Extraction Methodology

The methodology for the extraction of stability derivatives presented in Sect. 2 was validated using the 6:1 prolate spheroid. The added mass coefficients were compared with the theoretical calculation as well as with the simulation study by [16].

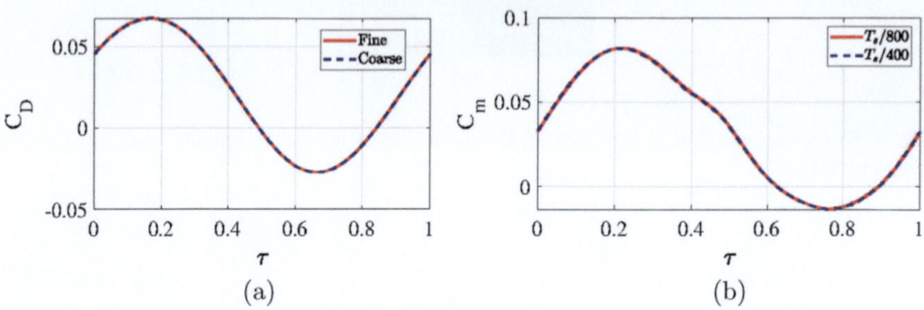

Fig. 6 **a** Grid independence and **b** time independence test results of the aerostat model

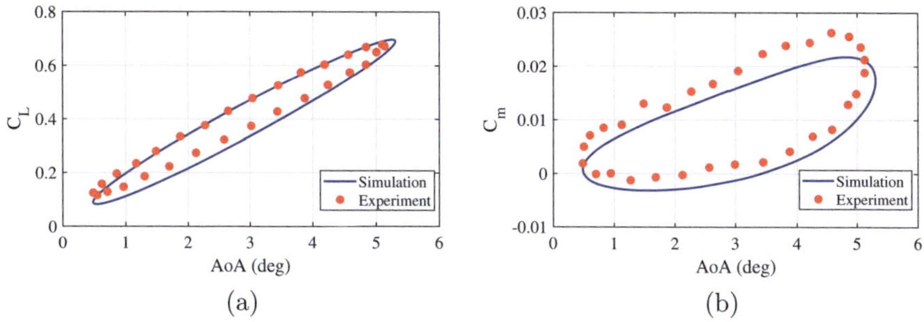

Fig. 7 Validation of grid motion; **a** lift coefficient and **b** moment coefficient comparison with AGARD CT1

For the estimation of $\dot{L}_{\dot{w}}$, a sinusoidal oscillation with amplitude 1 m/s and frequency 3.9 rad was used. For the estimation of $\dot{M}_{\dot{q}}$, two sinusoidal oscillations with amplitude 5° and frequencies $\omega_1 = 29.79$ rad and $\omega_2 = 21.06$ rad were used.

The theoretical calculations can be obtained using the following equations [33],

$$\dot{L}_{\dot{w}} = -\frac{4}{3}\pi\rho abc\frac{C_0}{2 - C_0} \tag{9}$$

$$\dot{M}_{\dot{q}} = -\frac{4}{15}\pi\rho\frac{abc(a^2 - c^2)^2(A_0 - C_0)}{2(c^2 - a^2) + (C_0 - A_0)(c^2 + a^2)} \tag{10}$$

The simulated value of $\dot{L}_{\dot{w}}$ is -0.0457 and the theoretical value is -0.0430, and for $\dot{M}_{\dot{q}}$, the corresponding values are -0.00339 and -0.0033.

5.4 *Acceleration and Frequency Independence Study*

As the current study involves the forced oscillation of the aerostat along the three degrees of freedom, the independence study of the stability derivatives from the amplitude of oscillations (acceleration of the motion) has to be done. Such a study was conducted for the Zhiyuan aerostat for heave oscillations, and the acceleration independence was tested for all the stability derivatives involved as shown in Fig. 8. The coefficients with almost the same values show the simulation's acceleration independence.

The estimation of moment stability derivatives involves the application of two sinusoidal oscillations with different frequencies. A frequency independence study was conducted for all the stability derivatives that required two frequencies. The results for the estimation of $\dot{M}_{\dot{q}}$ for a 10° angle of attack are shown in Fig. 9. Here, three frequencies are used for the comparison; 2.32 rad/s (2.7 s), 2.09 rad/s (3 s) and 1.90 rad/s (3.3 s). The frequency independence of the simulation is clearly observable in Fig. 9.

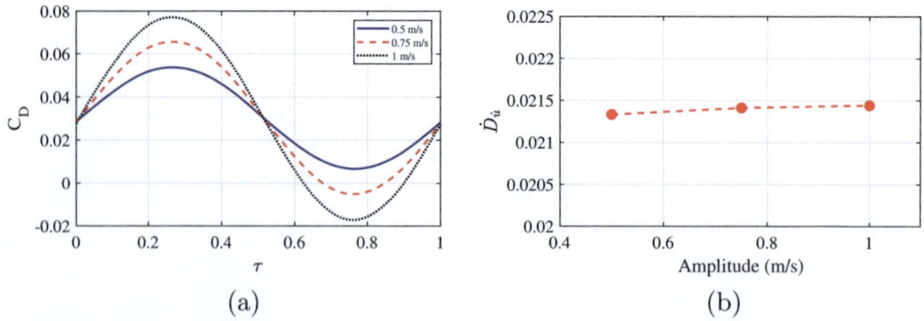

Fig. 8 Acceleration independence study results; **a** a full cycle drag response and **b** stability derivative, $\dot{D}_{\dot{u}}$, comparison

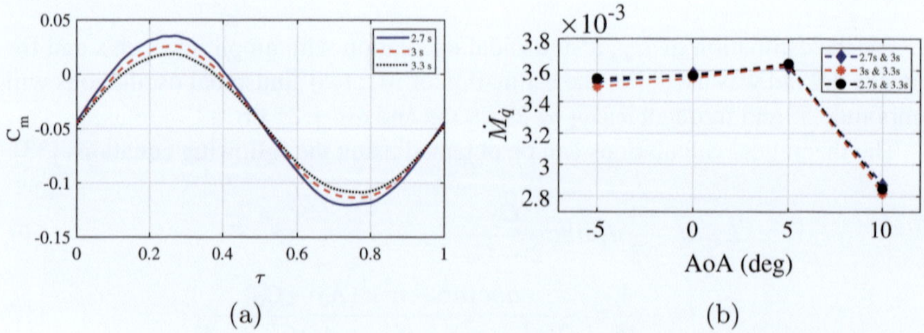

Fig. 9 Oscillation frequency independence study results; **a** a full cycle moment response and **b** stability derivative, $\dot{M}_{\dot{q}}$, comparison

The stability derivative extraction methodology presented in Sect. 2 was applied to the four aerostats with four different angles of attack. Three full cycles of responses were obtained for all the simulation cases, and a full stable cycle was selected for the analysis. The surge and heave oscillations were simulated using an amplitude of oscillation of 1 m/s. The drag, lift and moment response were measured for each aerostat shape for the heave oscillations. All the responses are plotted with the non-dimensional time on the X-axis.

5.5 Zhiyuan Aerostat

The response of the Zhiyuan aerostat for surge and heave oscillations is shown in Fig. 10. The figure shows the lift response for the heave oscillation. The first column of the figure represents a continuous set of responses and the second column represents the single-cycle plots that are used for the stability derivatives extraction. As the input oscillations are sinusoidal, the responses are also sinusoidal, as shown in the figure.

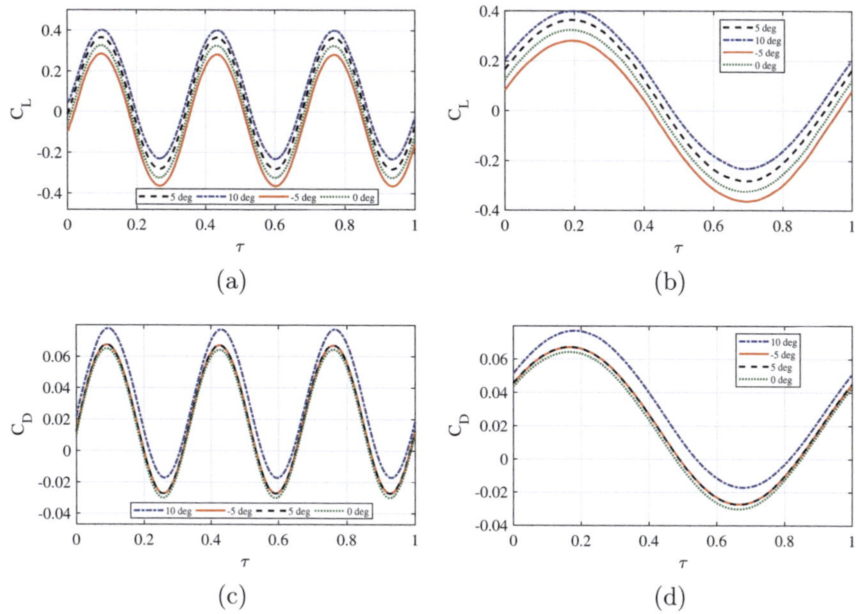

Fig. 10 Response of Zhiyuan aerostat for **a** heave oscillation-3 cycles, **b** heave oscillation-1 cycle, **c** surge oscillation-3 cycles and **d** surge oscillation-1 cycle

The magnitude of lift increases with the increase in the angle of attack, as shown in Fig. 10a. It is due to the variation in the differential pressure between the top and the bottom surfaces of the aerostat. Further increase in the angle of attack may cause stall condition, and it is not in the scope of the current study. The drag remains constant for the smaller angles of attack, as shown in Fig. 10c, because the frontal area hit by the wind remains almost unchanged. As the angle of attack increases to 10°, the frontal area of the aerostat increases and the drag increases.

The stability derivatives of the Zhiyuan aerostat due to the heave oscillation were extracted using the methodology specified in Sect. 2. A comparison of the stability derivatives extracted from the surge oscillations and heave oscillations for different angles of attack are shown in Figs. 11 and 12.

Drag derivative due to axial acceleration and velocity shown in Fig. 11a, b, respectively, shows almost constant values for the small angles of attack. As the angle of attack increases to 10°, the derivative increases as well. This is in close agreement with the drag response shown in Fig. 10a. Lift derivative due to the axial acceleration and velocity shown in Fig. 11c, d, respectively, shows a linear variation with the angle of attack. This shows the peculiar behaviour of the aerostat to build up the lift as the axial acceleration happens, which aids the dynamic stability of the aerostat. The magnitude of the derivatives shifts from negative to positive as the angle of attack changes from negative to positive. At the zero degree angle of attack, the derivative value is close to zero. That means the acceleration of the aerostat at

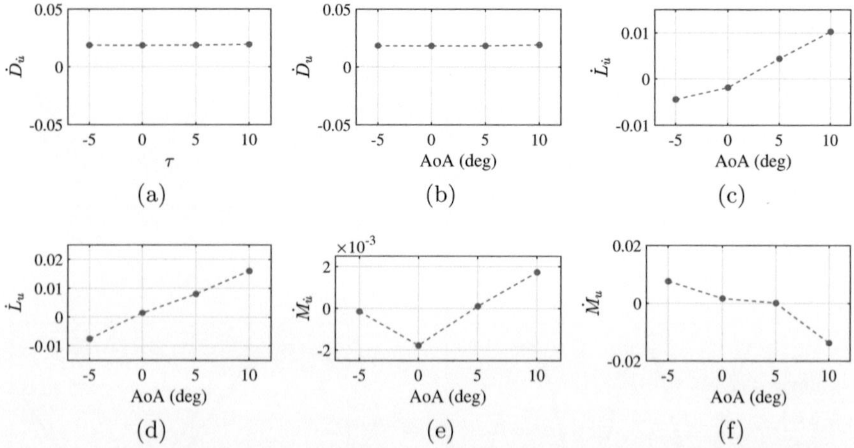

Fig. 11 Zhiyuan aerostat stability derivatives for surge oscillations at different angles of attack

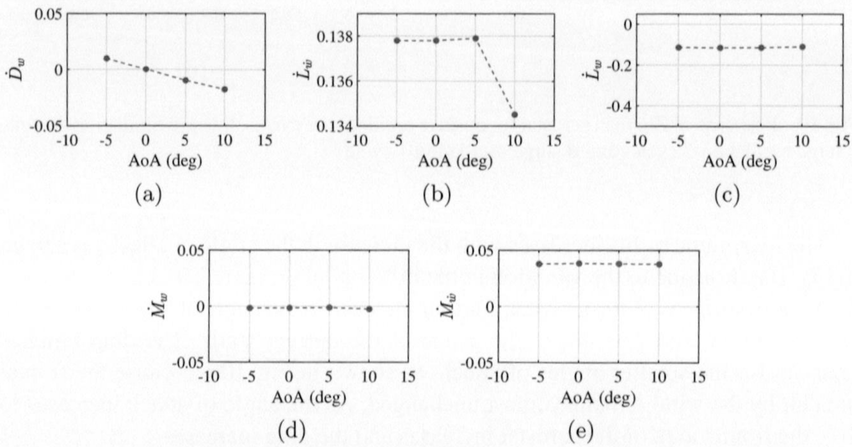

Fig. 12 Zhiyuan aerostat stability derivatives for heave oscillations at different angles of attack

zero degrees angle of attack will not affect much on the lift. Moment derivative due to axial acceleration and velocity shown in Figs. 11e, f, respectively, show a linear decrease with the angle of attack. There is an exception at a 10° angle of attack for the acceleration derivative. Thus, the acceleration along the axial direction causes the moment to decrease, reducing the control effort for pitch control.

Drag derivative due to the vertical velocity, shown in Fig. 12a, is linearly decreasing with the angle of attack. In contrast, the lift derivative is almost constant, as shown in Fig. 12c. The lift derivative due to vertical acceleration shown in Fig. 12b remains constant for small angles of attack but reduces to a lower value for the higher angle of attack. The vertical acceleration mainly caused by vertical gusts will not aid the drag build up of the aerostat. On the other hand, it will not even benefit the lift generation of the aerostat. The moment derivative due to vertical velocity shows a variation opposite to the moment derivative due to axial acceleration, as shown in Fig. 12c. The moment derivative due to vertical acceleration remains constant for all the angles of attack with an exception at zero degree, as shown in Fig. 12d.

5.6 GNVR, HAA and NPL Aerostats

The simulations for extracting the stability derivatives due to the surge and heave oscillations were repeated for the GNVR, HAA and NPL aerostats as well. Here in this section, some of those results which showed deviation from the Zhiyuan response are presented.

The response of the GNVR aerostat, HAA aerostat and NPL aerostat for surge and heave oscillations is shown in Fig. 13. A stable cycle of response used for the stability derivatives extraction is shown in the figure.

The lift response of the GNVR aerostat for different angles of attack shows an increment with the angle of attack as shown in Fig. 13a. The close similarity between the Zhiyuan shape and GNVR shape made the lift response of these two aerostats similar. The drag response of the GNVR aerostat shown in Fig. 13b has a considerable deviation from that of the Zhiyuan aerostat. Due to the more streamlined design, the Zhiyuan aerostat has a reduced drag compared to the GNVR aerostat. The lift response of the HAA aerostat for different angles of attack is shown in Fig. 13c. The amplitude of the lift increases with the increase in the angle of attack. The drag response shown in Fig. 13d demonstrates an invariant relation with the angle of attack. The moment response, shown in Fig. 13f, reduces with the increase in the angle of attack. The lift response of the NPL aerostat shown in Fig. 13e shows the least lift magnitude for the given angles of attack. The drag response remains constant for the smaller angles of attack and shows a slight increase in the magnitude for the 10° angle of attack, as shown in Fig. 13f. The drag magnitude is also the least among the four aerostats.

The stability derivatives obtained for the Zhiyuan aerostat is compared with those for the other aerostats, as shown in Fig. 14. The drag derivatives due to axial acceleration for the four aerostats shown in Fig. 14a remain constant with the angle of attack. Even though the individual derivatives remain constant, their magnitude varies largely among the aerostats. GNVR aerostat has the maximum value, and the NPL has the minimum value, as shown in the figure. As seen for the Zhiyuan aerostat case in Fig. 11a, the drag derivatives for the 10° angle of attack show a deviation for all the aerostats. It is due to the increase in the frontal area of the aerostat due to the large angle of attack. The magnitude of drag derivative for the GNVR aerostat is slightly higher than that for the Zhiyuan aerostat. This can be explained by the slightly larger diameter of the GNVR aerostat, which causes the front area to increase slightly. In a similar way, the diameter of the NPL aerostat is smaller than the Zhiyuan aerostat, which causes the front area to be smaller than the other aerostat, thereby resulting in a smaller drag. This drag magnitude pattern is reflected in the rate of change of drag

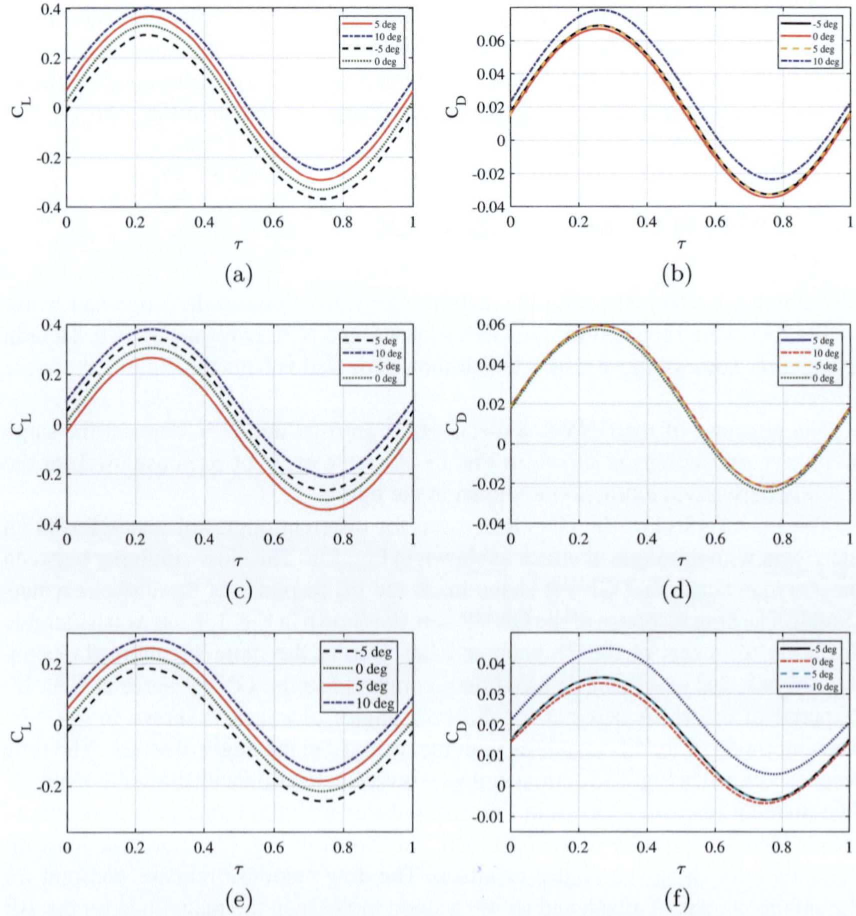

Fig. 13 Response of GNVR aerostat for **a** heave oscillation and **b** surge oscillation, HAA aerostat for **c** heave oscillation and **d** surge oscillation, and NPL aerostat for **e** heave oscillation and **f** surge oscillation

as well. As the diameter of Zhiyuan and HAA aerostats was the same, drag response and the drag derivative remained the same, as shown in the figure. The lower drag derivative value of the NPL aerostat helps to maintain the attitude of the aerostat because as the aerostat accelerates forward, the drag build up will be comparatively less among the four aerostats. On the other hand, the other three aerostats need more station-keeping control effort.

The lift derivative due to the axial acceleration for the four aerostats is shown in Fig. 14b. There is a peculiar difference between the derivative response of the Zhiyuan aerostat among the other aerostats. As the angle of attack increases, the rate of change of lift due to the axial acceleration increases linearly, whereas the derivatives decrease for the other aerostats. This particular behaviour of the Zhiyuan

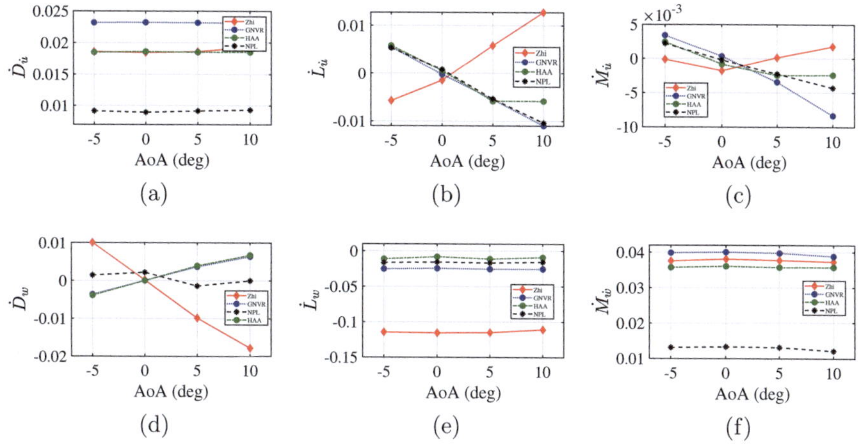

Fig. 14 Comparing the stability derivatives of Zhiyuan aerostat with other aerostats

aerostat is advantageous for the dynamic stability of the aerostat as the increased lift helps the aerostat to be more stable. The geometry of the Zhiyuan aerostat causes this particular response. All the other aerostats considered have a less pointy leading edge than the Zhiyuan aerostat. The positive slope in the lift derivative with the axial acceleration of the Zhiyuan aerostat aids the altitude control and station-keeping control efforts. The acceleration along the axial direction causes the lift to reduce its magnitude for the other three aerostats, unlike the Zhiyuan aerostat case. It shows that the aerostat fails to retain its lift as its own. There should be control actions to maintain the required lift for the aerostat.

The moment derivatives due to the axial acceleration for the four aerostats are shown in Fig. 14c. The rate of change of moment decreases with the angle of attack for the other three aerostats. The reduction in the moment derivative due to the axial acceleration will reduce the control effort for the pitch control. The Zhiyuan aerostat moment derivative increases for the positive angle of attack, as shown in the figure. Thus, the Zhiyuan aerostat demands more control effort for maintaining the desired pitch for large positive angles of attack. Even though the stalling of the aerostat is not in the scope of this work, the pattern in the moment derivative for the Zhiyuan aerostat suggests that stalling occurs at a lower angle of attack than for the other three aerostats.

The drag derivatives due to the vertical velocity for the four aerostats are shown in Fig. 14d. There is a positive slope for the derivatives plot for the GNVR, HAA and NPL aerostats and a negative slope for the Zhiyuan aerostat. Unlike the Zhiyuan aerostat, the drag magnitude builds up as the other three aerostats accelerate in the vertical direction, primarily due to vertical wind gusts. This demands more control effort for the station keeping and attitude control for the aerostats. This shows that as the turning of the aerostat for a pitch up flight, such as taking off from the ground station, slows down the aerostat due to the extra drag build up. The negative slope of the Zhiyuan aerostat aids the control efforts to maintain the position of the aerostat.

The lift derivatives due to the vertical velocity for the four aerostats shown in Fig. 14e possess constant values for different angles of attack. The vertical acceleration can be due to the presence of a strong vertical wind gust or an ascending flight phase at the time of the installation or maintenance of the aerostat. The Zhiyuan aerostat has the maximum lift derivative in the negative direction, which demands more control effort for the station keeping. All the other aerostats aid the lift for the ascending phase, which helps to reduce the control effort. The moment derivatives due to the vertical acceleration for the four aerostats shown in Fig. 14f also remain constant for different angles of attack. Zhiyuan, HAA and GNVR aerostats have the maximum value among the four aerostats, making the pitch control rapid and requiring more effort to keep them in the desired pitch. The NPL aerostat has the minimum value, which helps the aerostat to vary the pitch smoothly with less control effort.

6 Conclusions and Future Scope

The presented study has considered the translational motion of the aerostat in X (surge) and Y (heave) directions for the extraction of the stability derivative. A small-amplitude sinusoidal oscillation was used for the simulation of the aerostat motion. A methodology for the extraction of stability derivatives from the full cycle oscillation responses of the aerostat was presented with CFD simulation results. A base case aerostat shape was selected for the analysis, and three other aerostat shapes were considered for the comparison of results obtained from the analysis.

In this paper, the stability derivatives of the Zhiyuan aerostat at four different angles of attack were presented. The results of which were compared with the derivatives of the other three aerostats as well. The drag and lift performance of the Zhiyuan aerostat are superior among the four aerostats. This will help the aerostat to be controlled with less control effort during the station keeping and attitude control tasks. The NPL aerostat had better moment stability for the pitching acceleration. This will make the Zhiyuan aerostat demand more control effort for the pitching manoeuvre.

From this study, it can be concluded that the Zhiyuan aerostat can be used for the low-altitude wind measurement application. The more aerodynamic geometry of the Zhiyuan aerostat compared to the other three aerostats was the main reason for this peculiar performance. The dynamic stability analysed using the stability derivatives suggests the superiority of the Zhiyuan aerostat among the four aerostats. The drag and lift performance of the Zhiyuan aerostat are such that it aids the station keeping and attitude control efforts.

This study will be extended for the investigation of the stability derivatives due to surge and pitch motions. The longitudinal stability derivatives can be extracted from the aerostat oscillations involving the longitudinal motion variables and can be used for the longitudinal dynamic model of the aerostat. An optimization of the aerostat geometry for improving the dynamic stability can be considered as a future direction of this study.

References

1. Somasundaran N, Kayani V, Chandrasekhar R, Kottayil S (2019) Int J Autom Smart Technol 9(3). https://doi.org/10.5875/ausmt.v9i3.1912
2. Sengar S, Liu X (2020) J Ambient Intell Humaniz Comput 11. https://doi.org/10.1007/s12652-020-01866-7
3. Vanitha V, Raphel D, Resmi R (2019) Innovations in power and advanced computing technologies (i-PACT), vol 1, pp 1–4. https://doi.org/10.1109/i-PACT44901.2019.8960017
4. Azad K, Alam M, Uddin S (2012) Int J Adv Renew Energy Res 1:48
5. Allison S, Bai H, Jayaraman B (2020) Aerosp Sci Technol 98:105699. https://doi.org/10.1016/j.ast.2020.105699
6. Anoop S, Oruganti VRM (2020) Adv Sci Technol Eng Sys J 5(4):167. https://doi.org/10.25046/aj050420
7. Anoop S, Velamati RK, Oruganti VRM (2021) Aerosp Sci Technol 113:106684. https://doi.org/10.1016/j.ast.2021.106684
8. Nguyen K, Au LTK, Phan HV, Park HC (2021) Aerosp Sci Technol 119:107085. https://doi.org/10.1016/j.ast.2021.107085
9. Bykerk T, Verstraete D, Steelant J (2020) Aerosp Sci Technol 98:105709. https://doi.org/10.1016/j.ast.2020.105709
10. Jones SP, DeLaurier JD (1983) J Aircr 20(2):120. https://doi.org/10.2514/3.44840
11. Badesha S (1993) 10th lighter-than-air systems technology conference, pp 1–4. https://doi.org/10.2514/6.1993-4036
12. Ignatyev DI, Khrabrov AN (2015) Aerosp Sci Technol 41:106. https://doi.org/10.1016/j.ast.2014.12.017
13. Waqar A, Ali OA, Erwin S, Mohamed AJ (2017) Aircr Eng Aerosp Technol 89:174. https://doi.org/10.1108/AEAT-06-2015-0165
14. Tao Q, Tan TJ, Cha J, Yuan Y, Zhang F (2021) Unmanned Sys 09(01):73. https://doi.org/10.1142/S2301385021500060
15. Junior JLM, Santos JS, Morales MA, Goes LC, Stevanovic S, Santana RA (2019) AIAA Aviation Forum. https://doi.org/10.2514/6.2019-2982
16. Wang XL (2012) J Aircr 49(3):933. https://doi.org/10.2514/1.C031634
17. Ronch AD, Vallespin D, Ghoreyshi M, Badcock KJ (2012) AIAA J 50(2):470. https://doi.org/10.2514/1.J051304
18. Ronch AD (2012) On the calculation of dynamic derivatives using computational fluid dynamics. Ph.D. thesis, University of Liverpool
19. Carrión M, Biava M, Barakos GN, Stewart D (2017) J Aircr 54(4):1328. https://doi.org/10.2514/1.C033987
20. Lee SK, Joung TH, Cheo SJ, Jang TS, Lee JH (2011) Int J Nav Archit Ocean Eng 3(3):174. https://doi.org/10.2478/IJNAOE-2013-0060
21. Mader CA, Martins JRRA (2014) AIAA J 52(11):2533. https://doi.org/10.2514/1.J052922
22. Bykerk T, Verstraete D, Steelant J (2020) Aerosp Sci Technol 103:105883. https://doi.org/10.1016/j.ast.2020.105883
23. Muller L, Libsig M, Martinez B, Bastide M, Bidino D, Yannick B, Roy JC (2020) AIAA Aviation Forum. https://doi.org/10.2514/6.2020-2782
24. Javanmard E, Mansoorzadeh S, Mehr JA (2020) Ocean Eng 215:107857. https://doi.org/10.1016/j.oceaneng.2020.107857

25. Ghoreyshi M, Jirásek A, Cummings RM (2014) Prog Aerosp Sci 71:167. https://doi.org/10.1016/j.paerosci.2014.09.001
26. Kale S, Joshi P, Pant R (2005) AIAA 5th ATIO and 16th lighter-than-air system technologies and balloon systems conferences. https://doi.org/10.2514/6.2005-7442
27. Liang H, Zhu M, Guo X, Zheng Z (2012) 50h AIAA aerospace sciences meeting including the New Horizons forum and aerospace exposition. https://doi.org/10.2514/6.2012-1180
28. Yang Y, Xu X, Zhang B, Zheng W, Wang Y (2020) Aerosp Sci Technol 98:105664. https://doi.org/10.1016/j.ast.2019.105664
29. Wang QB, Chen JA, Fu GY, Duan DP (2009) J Zhejiang Univ Sci A Appl Phys Eng 10:1609. https://doi.org/10.1631/jzus.A0820814
30. Khoury G (2012) Airship technology. Cambridge University Press, Cambridge
31. Ansys (2020) Academic research FLUENT, release 20.2, Help system, fluent theory manual. Guide. ANSYS, Inc., Canonsburg, PA, USA
32. Compendium of unsteady aerodynamic measurements. AGARD-R-702. Report (Aug 1982)
33. Korotkin A (2009) Added masses of ship structures, vol 88. https://doi.org/10.1007/978-1-4020-9432-3

If you have any concerns about our products,
you can contact us on
ProductSafety@springernature.com

In case Publisher is established outside the EU,
the EU authorized representative is:
Springer Nature Customer Service Center GmbH
Europaplatz 3, 69115 Heidelberg, Germany

Printed by Libri Plureos GmbH
in Hamburg, Germany